A PRIMER ON INTEGRAL EQUATIONS OF THE FIRST KIND

THE PROBLEM OF DECONVOLUTION AND UNFOLDING

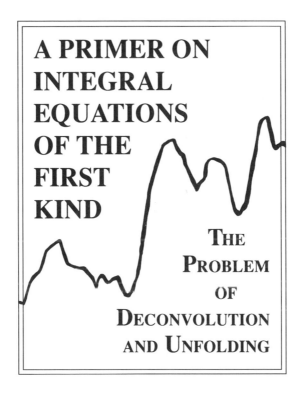

A PRIMER ON INTEGRAL EQUATIONS OF THE FIRST KIND

THE PROBLEM OF DECONVOLUTION AND UNFOLDING

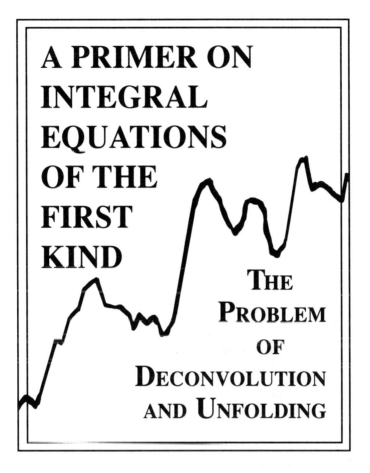

G. MILTON WING

with the assistance of
JOHN D. ZAHRT

LOS ALAMOS NATIONAL LABORATORY
LOS ALAMOS, NM

siam.

Society for Industrial and Applied Mathematics
Philadelphia

Library of Congress Cataloging-in-Publication Data

Wing, G. Milton (George Milton), 1923-
    A primer on integral equations of the first kind : the problem of deconvolution and unfolding / G. Milton Wing, with the assistance of John D. Zahrt.
        p. cm
    Includes bibliographical references and index.
    ISBN 0-89871-263-7
    1. Integral equations. I. Zahrt, John D. II. Title.
QA431.W644 1991   91-38059
515'.45—dc20

© 1991 Society for Industrial and Applied Mathematics

All rights reserved. No part of this book may be reproduced, stored, or transmitted in any manner without the written permission of the Publisher. For information, write to the Society for Industrial and Applied Mathematics, 3600 University City Science Center, Philadelphia, Pennsylvania 19104-2688.

Printed by Port City Press, Baltimore, Maryland.

In memory of Richard E. Bellman
—mentor, colleague, and friend.

Contents

Preface   ix

1 An Introduction to the Basic Problem   1
    1.1 The Problem Posed 1
    1.2 Interpretation of Integral Equations of the First Kind . . 3
    1.3 Some Ground Rules 3
    1.4 A Minor Difficulty Overcome 4
    1.5 Summary 4
    Problems I 5

2 Some Examples   7
    2.1 Introduction 7
    2.2 The Laplace Transform 7
    2.3 Some Remarks About Transport Theory 8
    2.4 Determination of a Signal by Absorption—Slab Geometry   9
    2.5 X-Ray Determination of the Density Distribution in a Sphere—PHERMEX Problem 11
    2.6 The Problem of Tomography 13
    2.7 Determination of Source Strength in an Absorbing Material 15
    2.8 Determination of the Radial Density Distribution of Plutonium in a Small Sphere 16
    2.9 A Problem in Geomagnetic Prospecting 17
    2.10 A Problem for the Department of Game and Fish 18

	2.11 A Problem in Mechanics	19
	2.12 Some Observations and an Important Concept	21
	2.13 Summary	22
	Problems II	22
3	**A Bit of Functional Analysis**	**27**
	3.1 Introduction	27
	3.2 Norms of Functions	27
	3.3 Norms of Vectors	29
	3.4 Norms of Integral Operators	30
	3.5 Norms of Differential Operators	32
	3.6 Norms of Matrix Operators	35
	3.7 Norms of General Operators T	36
	3.8 Summary	37
	Problems III	38
4	**Integral Operators with Separable Kernels**	**41**
	4.1 Introduction	41
	4.2 Separable Kernels	41
	4.3 Some Properties of Integral Operators with Separable Kernels	42
	4.4 IFKs with Separable Kernels	43
	4.5 Eigenvalues and Eigenfunctions of Integral Operators with Separable Kernels	45
	4.6 Integral Equations of the Second Kind with Separable Kernels	46
	4.7 Summary	48
	Problems IV	48
5	**Integral Operators with General Kernels**	**51**
	5.1 Introduction	51
	5.2 Some More Facts About Integral Operators—Existence and Boundedness of Inverses	52
	5.3 Eigenvalues and Eigenfunctions of Integral Operators with Symmetric Kernels	56
	5.4 Eigenvalues and Eigenfunctions of Nonsymmetric Operators—Singular Values and Singular Functions	58
	5.5 Integral Equations of the Second Kind with Symmetric Kernels	63

	5.6	Integral Equations of the First Kind with General Kernels	65
	5.7	Summary	66
	Problems V		67

6 Some Methods of Resolving Integral Equations of the First Kind — 71
6.1 Introduction — 71
6.2 Classical Quadrature Approach — 72
6.2.1 Basic Scheme — 72
6.2.2 A Particular Application of the Quadrature Approach — 73
6.2.3 A Trivial Example and Some Important Observations — 76
6.2.4 Section 6.2.2 Revisited — 78
6.2.5 A Return to the Kernel $K = 1 + \eta xy$ — 78
6.2.6 The Condition Number of a Matrix — 79
6.2.7 A Digression Concerning Approximate Kernels — 80
6.2.8 The Method of Least Squares — 81
6.2.9 Some Miscellaneous Comments Concerning the Quadrature Method — 83
6.2.10 A Summary of the Classical Quadrature Approach — 85
6.3 General Series Expansions — 85
6.3.1 Basic Scheme — 85
6.3.2 Series Expansions and Projection and Collocation Methods — 87
6.4 Expansion in Series of Eigenfunctions and Singular Functions — 89
6.4.1 Basic Scheme — 89
6.4.2 A Variant of the Method — 90
6.5 Parametrized Solution Function Method — 91
6.6 The Method of Regularization — 92
6.6.1 Some Background and the Basic Scheme — 92
6.6.2 Some Numerical Examples — 97
6.6.3 The Selection of the Parameter, γ — 103
6.6.4 A Generalization of the Concept of Regularization — 103
6.7 Stochastic Approaches — 103
6.8 Iterative Methods — 106
6.9 Summary — 107
Problems VI — 107

7 Some Important Miscellany — 113
- 7.1 Introduction — 113
- 7.2 About Those Ground Rules — 113
- 7.3 A Final Word About Codes and Quadrature Schemes — 114
- 7.4 Finding Generalized Moments of Solutions — 115
- 7.5 IFKs in More Variables — 116
- 7.6 Errors in the Kernel — 117
- 7.7 Summary — 117
- Problems VII — 117

8 Epilogue — 119

References — 121

Appendix A — 125

Appendix B — 127

Appendix C — 129

Index — 133

Preface

My first serious encounter with an integral equation of the first kind occurred about 1955 at the Los Alamos Scientific Laboratory, now renamed the Los Alamos National Laboratory. I was approached by a group of physicists who were building a very large x-ray machine and who desired some mathematical help. (Some detail is provided in Section 2.5 of this book.) The difficulties I encountered were very surprising to me.

Although my interest in integral equations of the first kind had been piqued, I did not resume research in the subject until over twenty years later. When I returned to Los Alamos in the early 1980s, I discovered vastly increased interest in such equations. The reason gradually became clear. Experimental techniques had long given rise to these equations, but they had been basically unmanageable. Now the computer provided an obvious way of attacking such problems.

But alas! They often remained unmanageable. A large collection of experimentalists found its way to my office. The code, which had worked satisfactorily for him or her before (or for someone else, who had kindly shared it) now produced total nonsense. What was wrong?

I found that my "clients" often had little mathematical training beyond "engineering advanced calculus." Many had never heard of an integral equation of the first kind, although frequently the terms "unfolding" and "deconvolution" were somewhat familiar.

After several years of experience with perplexed chemists, engineers, geologists, etc., I concluded that perhaps a Laboratory report on the subject would be useful. Moreover, that report should be written in a conversational style and assume the level of mathematical training I was regularly encountering. The material should be skewed toward Laboratory interests.

Eventually a document (LA-UR-84-1234) with the same title as that of this book emerged. Although it was intended primarily for internal Los Alamos National Laboratory use, it became more widely available through the Department of Energy distribution system. In time I received requests from other National Laboratories, from industry, from universities in this country, and then from overseas. About two years ago the Society for Industrial and Applied Mathematics approached me about the possibility of their publishing the report in somewhat revised form. This book is the result.

The publisher requested that the overall approach and style be retained. I gladly agreed. You, the reader, may at times find the "laid back" expositions somewhat patronizing or condescending. If so, I am sorry. I regret that I cannot adjust my level and tone for every reader as I can for my "clients."

If you have a better mathematical background than that assumed you may be distressed at the "sloppy" mathematical exposition, the lack of carefully stated (and proved) theorems, the informal definitions, etc. For you, I have given references to works which provide elegant and mathematically correct discussions. My observation of the reaction of many non-mathematicians to mere mention of "compact operators" or the "Lebesgue integral" has led me to conclude that a more traditional mathematical exposition would simply frighten off my primary intended audience.

Although the original Laboratory report has formed the basis of this book there are numerous changes and additions. The slant toward Laboratory interests has been at least partially eliminated. New topics and concepts have been added. More examples of all types, physical and numerical, have been included. An effort has been made to produce a more "textbook"-like work, suitable for either self-study or for the classroom.

Problems have been added. These are neither of the routine drill variety, nor of a research nature. Each problem is devised to illustrate an important topic covered. Virtually all of them should be attempted by you, the reader. Had space allowed, many would have been included as examples in the text. Some require the use of a computer. For the most part a relatively unsophisticated personal computer plus modest software will suffice. Do not feel reluctant to modify a problem if, as stated, it seems to exceed your equipment's capabilities.

You are perhaps anticipating lengthy descriptions of sophisticated numerical methods applied to special classes of problems plus associated codes. These you will not find. The properties of integral equations of the first kind are so unusual, unpleasant, and revealing that it seems appropriate to concentrate on those properties. An understanding of them is very helpful in eventually seeking "solutions." When numerical ideas are introduced they are discussed at the conceptual level, although some numerical experiments are described.

The word *solutions* used in the latter part of the preceding paragraph is enclosed in quotation marks. That is because in most instances a solution in the usual sense can seldom be found, not even a good approximate (numerical) solution. Integral equations of the first kind are quite different from differential equations. For those a solution of almost arbitrarily high accuracy may often be computed. Not so with our equations! In fact, I use the (nonstandard) term "resolve" instead of "solve" or "solve approximately" to emphasize the unusual difficulties. Of course, some classes of integral equations of the first kind are more easily resolved than others. I do not dwell on those classes, preferring to keep the overall discussion quite general. My primary goal is *basic understanding*.

Above all, it should be recognized that this book is a *primer*—an *elementary* textbook. There is probably very little here of interest to the expert, except for the possible potential use of the material for instructional purposes. The level of mathematical knowledge required is that implied earlier—namely, "engineering advanced calculus," which usually includes some elementary matrix theory and a small amount of numerical analysis. As I have already noted, the content and mode of presentation have been strongly influenced by my experiences at the Los Alamos National Laboratory. I hope that the resultant bias does not reduce the possible value of the book to others whose professional experiences may be somewhat different from mine.

I am indebted to a large number of people for making this overall effort possible. The original Laboratory report was carefully read and critiqued by several colleagues including Andrew B. White, Jr., Vance Faber, and Thomas A. Manteuffel. Numerous readers who have subsequently obtained the report have made helpful comments.

In revising the report into book form I have found the assistance of John D. Zahrt invaluable. He has always been readily available for lengthy critical discussions. John made many important suggestions, provided some numerical results, and aided with the proofreading. His Science and Engineering Research Semester student, Mark Kust, of Western Michigan University, obtained the bulk of the computational results.

The Los Alamos National Laboratory provided facilities, equipment, and personnel. Most important of these people was Yvonne Martinez. Yvonne did some of the wordprocessing of the original Laboratory report and all of the wordprocessing of the book itself. Her expertise, coupled with her ever-smiling demeanor (even when confronted with yet another "May I make just one more little change...?") have really made this part of my work quite pleasant.

Finally, my thanks to Vickie Kearn and the entire SIAM publishing staff.

Despite my best efforts I am sure errors will be found. I absolve from responsibility all of the above-mentioned people. *Mea culpa!*

Enough! Now let us "resolve"....

G. Milton Wing
Santa Fe, NM
May 1991

CHAPTER 1

An Introduction to the Basic Problem

1.1 The Problem Posed

Two of the most standard forms of linear integral equations are

(1.1) $$f(x) = g(x) + \gamma \int_a^b K(x,y)f(y)dy$$

and

(1.2) $$g(x) = \int_a^b K(x,y)f(y)dy.$$

Here, both the function $K(x,y)$, called the *kernel*, and the function $g(x)$ are known, K on the square $a \leq x, y \leq b$, and g on the interval $a \leq x \leq b$. The quantity γ is a given constant parameter. The function f is to be determined on $[a,b]$.

Equation (1.1) is a linear Fredholm equation of the *second* kind. For it, a well-established theory exists, and an ever-increasing collection of viable numerical methods is available for determining the unknown f approximately [5, 7, 8, 15, 16]. Books on integral equation theory are weighted heavily toward expositions of this type of equations [11, 33, 46]. Such equations have interesting properties. Many of these can be predicted by anyone familiar with linear algebra and matrix theory. (For excellent treatments of these subjects, see [20, 47].)

Equation (1.2) is a linear Fredholm integral equation of the *first* kind (IFK). It too has a well-developed theory, but often that theory is only touched upon in textbooks. (See, however, [33, 46].) It also has a large number of known properties. In contrast to those possessed by equations of the second kind, many of these properties are very unpleasant.

They are often surprising even to mathematicians. Moreover, quite a few of these characteristics make the understanding of such equations—especially at the intuitive level—quite difficult. There is a paucity of satisfactory numerical methods for determining the unknown function f.

I pause briefly to further explain my use in the Preface of the word "resolve." A solution to Eq. (1.2) can seldom be obtained in closed form. Frequently, in practical situations, a reliable approximation to the solution cannot even be found, a consequence of the unpleasant properties mentioned above. It is usually necessary to settle for less information than we would like. Thus, I choose to speak of resolving, rather than solving, such problems.[1]

If you have encountered equations of the form Eq. (1.2), you are perhaps familiar with the term "unfolding." Strictly speaking, this word applies only to equations of convolution, or folding, type ("folding" stemming from the German "Faltung"), but common usage has broadened the concept of unfolding to include the resolution of any Fredholm integral equation of the first kind. The word "deconvolution" is used in a similar way.

IFKs arise constantly in practice. In the next chapter, several examples are given. Many of these have come to my attention through work carried out at the Los Alamos National Laboratory. The preponderance of such examples is a reflection of my personal experience. It should not be concluded that atomic energy work is the primary source of IFKs. Experimental techniques in many fields (for example, remote sensing) give rise naturally to such equations. Subsequent attempts to interpret experimental results lead to the questions to be discussed in this book.

In this book, I identify and discuss the real source of the difficulties posed by IFKs, suggest methods for resolving such equations, and describe more precisely the distinction between *solution* and *resolution*. We shall emerge, it is hoped, with a greater respect for IFKs and their basic limitations.

We shall discover that numerical approaches are often required. When this is the case, the basic ideas will be introduced, but sophisticated questions of numerical analysis will not be discussed, nor will any codes be explicitly given. When necessary (for example, in problems), the nu-

---

[1] Sometimes the expression "inverting the IFK" is used. In practice this inversion is almost always approximate. I prefer the term "resolution" to "approximate inversion."

An Introduction to the Basic Problem

merical details must be supplied by you. Doubtless they will depend on your background, available computing facilities, etc. For the most part, relatively simple approaches will serve to demonstrate the points being made.

1.2 Interpretation of Integral Equations of the First Kind

In an experimental context, the function g in Eq. (1.2) often represents experimental data, obtainable only at a finite set of points $x = x_i$. Frequently these points are few, and the data usually contain errors. Sometimes data are irreproducible, as is the case with many experiments involving destructive testing.

The function K usually represents the response of the experimental equipment. This function can ordinarily be obtained through controlled laboratory experiments to a relatively high degree of accuracy and for many values of the variables (x, y). In certain instances, much is known about K analytically. Therefore we consider K as completely known.

Finally, the function f is the signal that is to be determined. Obviously, it is ordinarily desirable to obtain as much information about the signal as possible.

1.3 Some Ground Rules

An outline of our basic assumptions and ground rules is as follows:

- All functions, variables, and parameters are real.

- Any function given analytically is well behaved. The exact meaning of this will depend upon the context and upon the reader's background. The reader may think in terms of continuity, although sometimes piecewise continuity or mild unboundedness will be appropriate.

- Use of the Dirac delta function will *not* be allowed.

- Whenever we are working in an experimental context, g will be known at only a finite number of points, $x_1, x_2, \cdots, x_N$, where N is relatively small. The exact meaning of small will depend upon the context. Moreover, $g(x_i)$ can contain error.

- The function K is known exactly at all meaningful values of x and y.
- The limits of integration a and b are finite.

Any deviation from these ground rules will be expressly noted.

1.4 A Minor Difficulty Overcome

In many integral equations which arise in practice it is *not* true that $K(x,y)$ is defined on the square $a \leq x, y \leq b$. Rather, $a \leq y \leq b$, while $c \leq x \leq d$. (Examples are found in the following chapter.) It is shown in Appendix A that adroit variable transformations can always be made to convert the problem to one in which x and y vary over the same interval and so K is indeed defined on a square.

For purposes of exposition and to simplify the theory we shall always use the $a \leq x \leq b$, $a \leq y \leq b$ convention.

1.5 Summary

In this short chapter I have posed our fundamental problem, roughly indicated its possible origins, and listed the ground rules upon which discussions will be based. Note that there are many problems involving IFKs in a realistic setting where our ground rules do not apply. For instance, in the general area of image restoration, the assumption of sparse data is inappropriate. Some of our work applies to such situations, but they are not foremost in my thoughts.

The assumption that K is fully known is often not completely justified; however, if K were also treated as in error, our problem would simply become more unmanageable. The matter is discussed briefly in Section 7.6.

Finally, many important *nonlinear* IFKs arise in practice. Such equations have the form

$$(1.3) \qquad g(x) = \int_a^b \tilde{K}(x,y,f(y))dy,$$

where $\tilde{K}$ cannot be written as $K(x,y)f(y)$. Even the theory for such equations is not completely understood and practical matters are still more complicated. Because the linear case itself is so difficult, nonlinear equations will not be discussed.

Problems I

1. Try to think of as many integral equations arising in your particular field of interest as you can. In each case, identify g, K, and f. Which equations are linear? Classify each of those as being of first or second kind.

2. You probably have encountered Green's functions, either in the study of differential equations or in physical applications. Consider the relationship between integral equations and problem formulations involving Green's functions.

CHAPTER 2

Some Examples

2.1 Introduction

This chapter is devoted almost entirely to a presentation of some examples of IFKs. As noted in the preceding chapter, many of these have arisen in work at the Los Alamos National Laboratory. Background material, discussion, and references (when available) are provided for these examples, but none is dealt with in any great depth. In those instances where an analytic solution to the equation is known, it is given. In every case this solution exhibits unpleasant behavior, indicating that IFKs may indeed be inherently troublesome.

2.2 The Laplace Transform

Although the Laplace transform [56] itself seldom occurs directly in the interpretation of experimental data, such interpretations do frequently lead to the transform. The inversion of the Laplace transform is really equivalent to the solution of an IFK. Let

(2.1) $$g(x) = \int_0^\infty e^{-xy} f(y) dy,$$

where the function g is known and f is to be determined. The kernel of this IFK is $K(x,y) = e^{-xy}$, where $a = 0$ and $b = \infty$. (This value of b represents a temporary suspension of the last of our ground rules.)

If g has been obtained experimentally or computationally, it is known at only a finite number of points, and there with error. You may recall the complex inversion formula for the recovery of f:

$$f(y) = \frac{1}{2\pi i} \int_{c-i\infty}^{c+i\infty} e^{yx} g(x) dx. \tag{2.2a}$$

Use of this formula is ordinarily precluded because it obviously requires information about the measured or computed quantity g in the complex plane.

A little-known but much more revealing formula has been derived by Post and Widder [56].

$$f(y) = \lim_{k \to \infty} \left\{ \left[\frac{(-1)^k}{k!} g^{(k)} \left(\frac{k}{y} \right) \right] \left(\frac{k}{y} \right)^{k+1} \right\}. \tag{2.2b}$$

Equation (2.2b) reveals that the inverse transform really depends upon the behavior of the kth derivative of g, as k becomes large. In addition, this behavior is more important for increasingly large arguments, k/y. Clearly, with data for g available at only a finite number of points—and there with inaccuracies—there is no hope of obtaining information concerning arbitrarily high derivatives.

There is a tendency to feel that Eq. (2.2a) therefore must be preferable to Eq. (2.2b) because no knowledge of derivatives seems to be required. If you are familiar with the properties of analytic functions, you will immediately note that this argument is specious. In fact, the overall problem of solving an IFK is not dependent upon the representation of the solution of that IFK, even when one or more explicit representations are available. The difficulty is inherent in the IFK. We return to this point later.

Although the Post–Widder formula demonstrates the gravity of the problem, it must be noted that there are numerous devices for numerically inverting Laplace transforms. No single one of these is effective in all cases. The algorithm used must be tailored to the problem at hand.

2.3 Some Remarks About Transport Theory

Several of the problems to be discussed involve concepts of transport theory, a subject with which you may not be familiar. Let me describe some basic concepts. For more information see, e.g., [57].

Consider a particle—say, a neutron—moving in a medium. It moves in a straight line until it suffers a collision with a nucleus of the material. The result of this interaction may be that the neutron disappears

Some Examples   9

(absorption), is scattered in a new direction (scattering), or creates new neutrons (fission). In this work we shall be primarily interested in absorption.

Obviously, an important consideration is the probability of a collision. It can be shown by fairly elementary arguments (see Problem 1 at the end of this chapter) that the probability of traveling distance x with *no* collision is

$$p(x) = e^{-\sigma x},$$

where σ, called the cross section, depends upon the medium, the energy of the neutron, and perhaps other physical parameters.

Very roughly speaking, σ depends upon the effective target area the moving particle "sees" as it moves through the material. It is not surprising that σ increases when the material is compressed (there are more targets in a unit length). Also one material may have "larger" targets than another. The problem of determining the cross section is a very complicated theoretical and experimental one which need not trouble us. In general, σ can be assumed to be known in our work.

Although we have talked of neutrons, similar remarks, with appropriate changes in terminology, may be made for photons, x rays, etc.

2.4 Determination of a Signal by Absorption—Slab Geometry

Many experiments are of the following general type. A signal in the form of a beam of particles is transmitted through a slab of material that absorbs some of the particles, but does not scatter any. The known absorption cross section depends upon the particle energy $\sigma = \sigma(E)$. The output of the experiment is a single number provided by a detector that merely measures the total number of particles striking it. The problem is to determine the energy dependence of the incident particle beam (see Fig. 2.1).

Mathematically this problem is easily formulated. If the slab thickness is x and the known cross section is $\sigma(E)$, then the detector measurement g is given by an equation of the form

(2.3) $$g(x) = \int_{E_{\min}}^{E_{\max}} e^{-\sigma(E)x} f(E) dE,$$

where $f(E)$ is the density of particles moving with energy E. It is obvious that it is impossible for a single experiment to reveal the desired energy

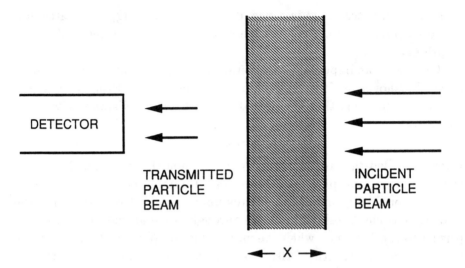

Figure 2.1. *Typical energy spectrum experiment.*

spectrum f. However, if the experiment is done for various thicknesses x, then Eq. (2.3) is simply an IFK with $K(x, E) = e^{-\sigma(E)x}, a = E_{\min}$, $b = E_{\max}$. (Note that the physics suggests that the variable y of Eq. (1.2) be replaced by E in this context.)

There is a close connection between this problem and the inversion of the Laplace transform. Suppose, in fact, that $\sigma(E)$ is monotonically increasing. We can then write

$$\sigma(E) = y, \qquad E = \sigma^{-1}(y),$$

$$\sigma(E_{\min}) = a, \qquad \sigma(E_{\max}) = b,$$

so that Eq. (2.3) becomes

$$g(x) = \int_a^b e^{-xy} f(\sigma^{-1}(y)) \frac{d}{dy} \sigma^{-1}(y) dy.$$

If it should also happen that $a = 0$ and $b = \infty$, then we have

(2.4) $$g(x) = \int_0^\infty e^{-xy} F(y) dy,$$

$$F(y) = f(\sigma^{-1}(y)) \frac{d}{dy} \sigma^{-1}(y).$$

Some Examples

This is the Laplace transform IFK.

When Eq. (2.3) is not equivalent to Eq. (2.4), the practical resolution of this problem still presents great difficulties. However, there are usually no analytic equations similar to Eq. (2.2b) to demonstrate this.

2.5 X-Ray Determination of the Density Distribution in a Sphere—PHERMEX[1] Problem

Let us assume a sphere of radius R made of a single material. The density of that material varies in an unknown fashion except that it depends only upon the distance from the center of the sphere. A plane-parallel beam of x rays is transmitted through the sphere and the attenuated beam collides with a photographic plate (see Fig. 2.2). Assuming no scattering or fission, can we determine the density distribution in the sphere from the darkening of the photographic plate?

After applying some relatively simple physical arguments, we find an equation of the form[2]

$$(2.5) \qquad I(z) = \exp\left\{-2\int_0^{\sqrt{R^2-z^2}} \sigma\left(\sqrt{z^2+s^2}\right) ds\right\}.$$

Here $I(z)$ is the intensity of the x rays at position z and $\sigma(r)$ is the cross section of the material at radial position r. We seek σ, from which the density can be calculated. (See [10, 36].)

Clearly, Eq. (2.5) is not of the standard form Eq. (1.2). However, a set of fairly easy mathematical substitutions transforms Eq. (2.5) into

$$(2.6) \qquad J(x) = \int_0^x \frac{\Sigma(y)}{\sqrt{x-y}} dy, \qquad 0 \le x \le R^2.$$

Here the function J is directly related to I, and Σ is dependent only upon σ; that is, knowledge of I gives J and knowledge of Σ yields σ and hence the desired density. (See Problem 5.)

Now Eq. (2.6) is of the form of our standard IFK with J replacing g, Σ replacing f, $a = 0$, $b = R^2$, and

$$K(x,y) = \begin{cases} \dfrac{1}{\sqrt{x-y}}, & 0 \le y < x, \\ 0, & x < y \le R^2. \end{cases}$$

---

[1] PHERMEX is a very large x-ray machine at Los Alamos National Laboratory.
[2] We shall always normalize in such a way that "stray" constants are unity.

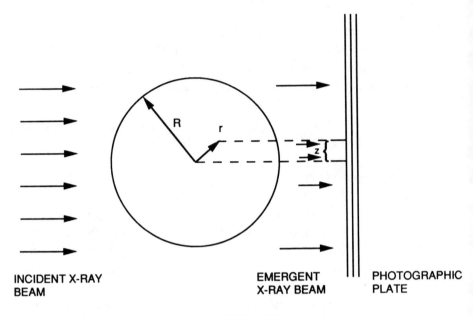

Figure 2.2. *Schematic of PHERMEX experiment.*

(It must be noted that K can no longer be directly interpreted as the response of a detector.)

Equation (2.6) is a special case of a Volterra equation whose solution was found by Abel [55]:

$$\begin{aligned}\Sigma(y) &= \frac{1}{\pi}\frac{d}{dy}\int_0^y \frac{J(x)}{\sqrt{y-x}}dx \\ &= \frac{1}{\pi}\left\{\frac{J(0)}{\sqrt{y}} + \int_0^y (y-x)^{-1/2}J'(x)dx\right\}.\end{aligned}$$

(2.7)

We observe again the presence of a derivative—although only a first derivative in contrast to the case in the Post–Widder formula Eq. (2.2b)—but J is experimentally determined and is known only inaccurately at a finite number of points. Numerical differentiation of such data is always delicate at best. Small errors in the measured quantity can produce large errors in the derivative approximations and can lead to large errors in Σ, and hence in σ.

2.6 The Problem of Tomography

A little thought reveals that the preceding problem is closely connected with modern tomography, as used, for example, in medical applications. If a patient's head were spherical and the skull contents were radially distributed, the foregoing would provide a noninvasive technique for studying the brain. It is mathematically unpleasant—although anatomically and aesthetically desirable—that the head, and most parts of the body, are not perfect spheres.

It is interesting to note that the PHERMEX work of the previous section was done in the mid 1950s, while the intensive study of medical tomography did not begin until over half a decade later (see [10, 12–14, 36]).

Consider a slice S of a body organ, as shown in Fig. 2.3. Assume only that σ (say for x rays) is known for the various bones and tissues that may lie in this organ section. We wish to find the composition of S.

Enclose S in a circle C. The organ section may now be considered as the interior of C, with cross section zero where there is no organ material. Introduce coordinates as shown and observe that the x-ray path is completely determined by ρ and ψ. After some geometry and physics, the following analogue to Eq. 2.5 is found:

$$(2.8) \quad \log I(\rho, \psi) = \int_\rho^R \left\{ \sigma\left(r, \psi - \cos^{-1}\left(\frac{\rho}{r}\right)\right) + \sigma\left(r, \psi + \cos^{-1}\left(\frac{\rho}{r}\right)\right) \right\} \frac{r\, dr}{\sqrt{r^2 - \rho^2}}.$$

Clearly, Eq. (2.8) is much more complicated than the IFKs which we have previously encountered. Let us expand the logarithm of the known data and the unknown cross section in Fourier series.

$$(2.9a) \quad \log I(\rho, \psi) = I_0(\rho) + \sum_{k=1}^{\infty} \left\{ I_{1,k}(\rho) \cos k\psi + I_{2,k}(\rho) \sin k\psi \right\},$$

$$(2.9b) \quad \sigma(r, \phi) = \sigma_0(r) + \sum_{k=1}^{\infty} \left\{ \sigma_{1,k}(r) \cos k\phi + \sigma_{2,k}(r) \sin k\phi \right\}.$$

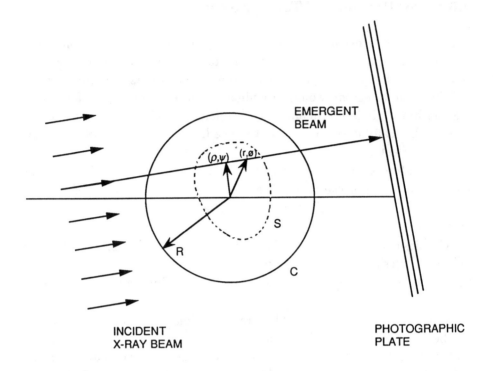

Figure 2.3. *Coordinate scheme for general tomography problem.*

Formal substitution of Eq. (2.9) into Eq. (2.8) and some knowledge of orthogonal polynomials eventually yield

$$I_{i,k}(\rho) = 2\int_{\rho}^{R} T_k\left(\frac{\rho}{r}\right)\frac{r}{\sqrt{\rho^2 - r^2}}\sigma_{i,k}(r)dr,$$

(2.10) $\quad\quad i = 1, 2, \quad k = 0, 1, 2, ...$

Equation (2.10) is an infinite set of IFKs for the unknown $\sigma_{i,k}$ in terms of the known coefficients $I_{i,k}$. The somewhat mysterious T_k functions are Chebyshev polynomials, whose mathematical properties are known (see [1] and Problem 14).

Thus this fairly general problem in tomography is reduced to the resolution of infinitely many IFKs of the type we have been discussing. (Many other approaches to tomography are known. See [30, 37, 40, 45, 52].)

2.7 Determination of Source Strength in an Absorbing Material

Let us begin with a very simple model. Suppose that a one-dimensional extent of material lies on the interval $(0,b)$, and that there are sources of neutrons distributed along this "rod." These neutrons can move only in two directions—right and left—along the rod, with the absorption cross section σ a known constant. An absorption deposits energy, and we are able to measure this energy. We wish to determine the location and strength of the sources from the energy deposition measurements.

Some elementary arguments (see Problem 7) lead to the equation

$$(2.11) \qquad g(x) = \int_0^b e^{-\sigma|x-y|} f(y) dy,$$

where g is a measure of the energy deposition at x and f is the source strength at y.

An analytical solution of this IFK is obtainable:

$$f(y) = \frac{1}{2}\left\{\sigma g(y) - \frac{1}{\sigma}g''(y)\right\}.$$

In this case, then, the desired function f depends upon the second derivative of the measured quantity. Numerically this is a somewhat worse situation than that discovered in Section 2.5.

This problem is obviously highly idealized. Next, consider a slab of material, of thickness b but infinite in extent, made of uniform, completely absorbing material with particle sources which depend only on the distance from the left face (Fig. 2.4). Let the particles emerge from the sources isotropically (that is, uniformly distributed in angle). The resulting integral equation is now

$$(2.12) \qquad y(x) = \int_0^b E_1(\sigma|x-y|) f(y) dy,$$

where

$$E_1(t) = \int_1^\infty \frac{e^{-tw}}{w} dw, \qquad t > 0.$$

An analytic solution to Eq. (2.12) is not known.

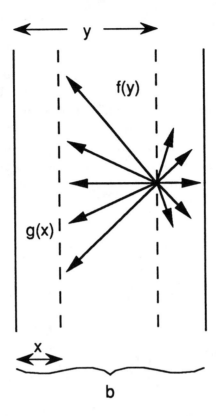

Figure 2.4. *Isotropic source in slab.*

2.8 Determination of the Radial Density Distribution of Plutonium in a Small Sphere

The problem referred to in this section's title arises in the investigation of the deposition of radioactive material in the lungs, and it has been studied by Hendry [29]. To be brief, we simply write the equation derived by him,

$$g(x) = \int_0^x \left[1 + \left(\frac{2}{x^2} - 1 \right) y - \frac{3y^2}{x^2} + \frac{y^3}{x^2} \right] f(y) dy.$$

Here $a = 0$, $b = 1$, (see Eq. (1.2)) and

$$K(x,y) = \begin{cases} 1 + \left(\frac{2}{x^2} - 1 \right) y - \frac{3y^2}{x^2} + \frac{y^3}{x^2} , & 0 \leq y \leq x, \\ 0 , & x < y \leq 1. \end{cases}$$

Some Examples

Again, the function g is obtainable only through measurement. The solution f is given by

$$f(y) = Ae^{-y} + \frac{1}{1-y}\left\{\frac{yg'(y)}{2} + g(y)\right\} - \frac{g(0)e^{-y}}{1-y}$$
$$- \frac{e^{-y}}{1-y}\int_0^y e^t\left[\frac{tg'(t)}{2} + g(t)\right]dt.$$

The value of the constant A is known but it is of no great interest here. What is of interest is the appearance again of the derivative of the data, g'. Hendry [29] makes no explicit use of this analytical solution.

2.9 A Problem in Geomagnetic Prospecting

For another example, we turn to an entirely different field. Suppose it has been determined (for example, through drilling) that there is a large ore deposit in a plane stratum at a known distance h beneath the surface of the earth. The deposit produces a measurable magnetic field at the surface, the intensity of which is an indication of the richness of the ore. Through magnetometry, we wish to determine where the most valuable deposits are located.

We consider a simplified one-dimensional model. Let the x axis represent the surface of the earth and the line $y = -h$ the position of the ore stratum. Suppose at $(s, -h)$ (see Fig. 2.5) the vertical component of the magnetic field has magnitude $m_y(s)$. The function $m_y(s)$ is what we wish to find. We are able only to measure $H_y(x)$, the vertical component of the field at the earth's surface. Now at P the value of H_y due to the part ds is

$$\frac{\sin\theta\, m_y(s)ds}{\left\{\sqrt{h^2 + (x-s)^2}\right\}^2} = \frac{hm_y(s)ds}{\{h^2 + (x-s)^2\}^{3/2}}.$$

If the deposit runs from $s = 0$ to $s = b$

(2.13) $$H_y(x) = h\int_0^b \frac{m_y(s)ds}{\{h^2 + (x-s)^2\}^{3/2}}.$$

This is an IFK for $m_y(s)$.

A somewhat more complicated problem results if we assume the richness of the ore known, but that the deposit lies on a nonplanar surface. Thus $m_y(s)$ is known in the model, but the straight line $y = -h$ must

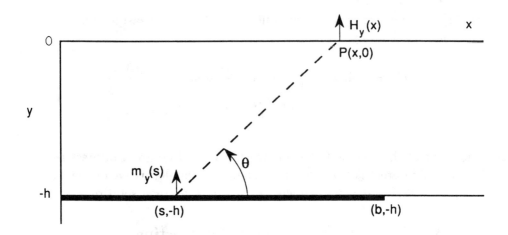

Figure 2.5. *The geomagnetic prospecting model.*

be replaced by an (unknown) curve. The analysis leads to a nonlinear integral equation. Such problems lie outside the scope of this book (see Section 1.5).

2.10 A Problem for the Department of Game and Fish

Quite commonly a new lake is created by damming up a stream. The local Department of Game and Fish then introduces an appropriate species of fish, ultimately for recreational purposes, sometimes closing the water to fishing for the first few years. It is valuable to learn how well the fish population is propagating.

Suppose the lake originally has no fish. It is stocked initially with a known number of trout of various sizes. Subsequent stocking is of fry (small trout) at the known rate $s(t)$ per year. The propagation rate $r(t)$ is unknown. At regular intervals the total number of trout $N(t)$ is estimated by selective netting. Of course, some fish die through natural causes. Often the mortality rate can be assumed to be simple exponential $e^{-\lambda t}$, where λ can be estimated from knowledge of trout in similar waters. We wish to determine $r(t)$.

The equation which applies in this situation is (see Problem 11)

(2.14) $$N(t) = N(0)e^{-\lambda t} + \int_0^t [r(y) + s(y)]e^{-\lambda(t-y)}\,dy.$$

Recall that $N(t)$ is known for $t \geq 0$, as is s. Hence Eq. (2.14) becomes

(2.15) $$g(t) = \int_0^t r(y) e^{-\lambda(t-y)} dy,$$

where

$$g(t) = N(t) - N(0)e^{-\lambda t} - e^{-\lambda t} \int_0^t s(y) e^{\lambda y} dy.$$

Equation (2.15) is an IFK for r. It can easily be solved. Write

(2.16) $$g(t) e^{\lambda t} = \int_0^t r(y) e^{\lambda y} dy.$$

Differentiation with respect to t and a bit of manipulation yields

(2.17) $$r(t) = g'(t) + \lambda g(t).$$

This result is deceptively easy. The function $g(t)$ involves $N(t)$, which contains large experimental error. Thus $g'(t)$ is probably very poorly estimated.

2.11 A Problem in Mechanics

As a final example we turn to a very classical problem, the tautochrone. Let a smooth wire be placed in a vertical plane, its lowest point at the origin O (Fig. 2.6). Suppose a bead slides down the wire under gravity and without friction. Can the wire be so shaped that regardless of which point $P(x, y)$ on the curve the bead starts from at rest, it reaches O in the same time T?

Let arc length on the curve, measured from O, be denoted by s. Consider the kinetic and potential energies of the particle as it passes (ξ, η) on the curve. By the standard conservation argument

(2.18) $$\frac{1}{2} m \left(\frac{ds}{dt}\right)^2 = mg(y - \eta),$$

where m is the mass of the bead and g is the acceleration of gravity. Thus

(2.19) $$ds = -\sqrt{2g} \sqrt{y - \eta}\, dt,$$

and so

(2.20) $$T\sqrt{2g} = \int_{\eta=0}^{\eta=y} \frac{ds}{\sqrt{y - \eta}}.$$

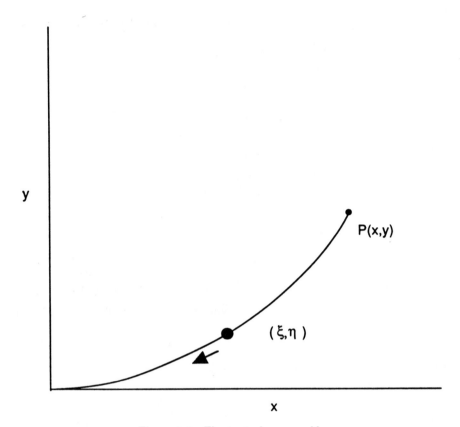

Figure 2.6. *The tautochrone problem.*

Now write
(2.21) $$s = H(y),$$
where H is determined by the curve. Thus

(2.22) $$T\sqrt{2g} = \int_0^y \frac{H'(\eta)d\eta}{\sqrt{y-\eta}}.$$

Upon proper identification of variables, Eq. (2.22) is recognizable as an Abel equation, which was encountered in Section 2.5. Thus H' can be found explicitly and the curve then determined. It is a cycloid! (See Problem 12.)

In this case the left-hand side of Eq. (2.22) is specified exactly. The data errors which have been so distressing in other problems do not exist. However, suppose we change the problem a bit. The wire is in place, but not visible to the experimenter. As assistant releases the bead on command, announcing the height y. The experimenter records the time of arrival $\tilde{T}(y)$ for many y values. Obviously, $\tilde{T}(y)$ will contain error. How accurately can the shape of the wire be determined? (See Problem 13.)

2.12 Some Observations and an Important Concept

For several of the examples presented in this section, no analytic solution is obtainable. Let us concentrate on the cases which have such solutions (Sections 2.2, 2.5, 2.7, 2.8, 2.10, 2.11).

In each of these examples, a *unique* solution has been found. Often it is argued that IFKs are difficult to handle because there is no unique solution. Sometimes the solution is not unique (see Problem 3). But in the sections cited, that is not the source of trouble. Rather, in the analytic solution presented in each case, a derivative of the data function has appeared. A small change in the data can result in a very large change in the derivative, and hence in the problem solution.

Compare this situation with one with which you are probably more familiar: the solution of differential equations. There small changes in the given functions or the boundary conditions (or any of their derivatives) *usually* have little effect on the solution to the problem. (There are many formal theorems of this kind.) With most IFKs this is *not* the case.

Observe also that the difficulties revealed are not in any way the result of numerical approximations. There have been no such approximations.

We might refer to IFKs as being "unstable" to such changes. Unfortunately, the word *unstable* has many meanings in mathematics and the natural sciences. Hadamard [26] introduced the concept of *well-posed* and *ill-posed* problems. A problem P is well posed if

(i) P has a unique solution, and

(ii) small changes in the data result in only small changes in the solution.

If either condition is violated, the problem P is ill posed.

In all the examples cited small changes in the data can cause large changes in the solution. These problems are thus ill posed. We shall see that IFKs are usually ill posed.

Many problems in applied mathematics share this property of ill posedness. Many cannot be phrased as IFKs. This general class of problems is under intensive investigation. To try to study the entire class would carry us afield. Our discussion will be confined to IFKs.

Our investigations have hence led, tentatively, to the concept of ill posedness. In the chapters which follow we shall try to better understand this phenomenon, and to determine what, if anything, can be done about it.

2.13 Summary

In this chapter, I have presented several examples of IFKs, including some problems that are, or have been, of direct interest and practical value to Los Alamos National Laboratory. Whenever a solution has been achieved, that solution has involved differentiation of the known function, that is, of the measured data. Numerical differentiation of discrete and inaccurate data is known to create problems. It has been indicated that similar difficulties occur in problems for which we are unable to provide an analytical solution. We investigate this matter in subsequent chapters.

Problems II

1. Try to obtain the probability (see Section 2.3) $p(x) = e^{-\sigma x}$ by the following argument. Let $p(x)$ be the probability that the particle moves distance x with no collision. Now suppose it moves an additional distance Δ without collision. Then $p(x + \Delta) = p(x)p(\Delta)$. It is reasonable to expect that the probability of a collision in a *very small* distance Δ is proportional to Δ. Thus $p(\Delta) \doteq 1 - k\Delta$. Obtain a differential equation for $p(x)$. What is k?

2. Obtain Eq. (2.3) choosing units in such a way that there are no "stray" constants.

3. Suppose that $\sigma(E)$ looks as shown in Fig. 2.7. (Such cross sections, which occur in the physical world, are said to have "edges.") Make an argument that there may be two or more signals f which yield the same g. Thus, for such a $\sigma(E)$, Eq. (2.3) does not have a unique solution.

Some Examples

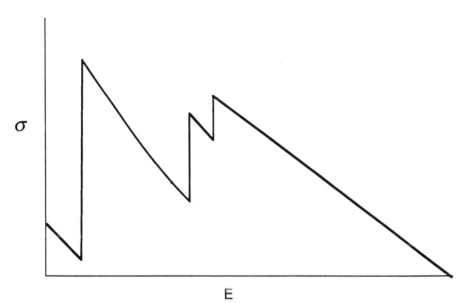

Figure 2.7. *Cross section with edges.*

4. By using geometrical arguments, obtain Eq. (2.5).

5. Make the substitutions required to obtain Eq. (2.6) from Eq. (2.5). Thus find the explicit relationship of J to I and of Σ to σ.

6. The Abel equation and Eq. (2.15) of Section 2.10 are both of the form

    $$(*) \qquad g(x) = \int_0^x K(x-y) f(y) \, dy.$$

    Equation $(*)$ is called a *convolution* equation. If you are familiar with the Laplace transform and the convolution theorem, formally solve Eq. $(*)$. Discuss any difficulties you perceive.

7. Using $p(x) = e^{-\sigma x}$, obtain Eq. (2.11) by physical reasoning. If you are acquainted with transport theory, obtain Eq. (2.12).

8. By careful differentiation obtain the solution to Eq. (2.11). (Hint: write

    $$\int_0^b e^{-\sigma|x-y|} f(y) \, dy = \int_0^x e^{-\sigma(x-y)} f(y) \, dy + \int_x^b e^{-\sigma(y-x)} f(y) \, dy$$

    and then differentiate.)

9. Using the fact that $T_0(x) \equiv 1$, show that the results of Section 2.6 reduce to that of Section 2.5 when I and σ are independent of angle.

10. Use Eq. (2.2b) to find the inverse Laplace transforms of $g(y) = 1/y^2$ and $g(y) = 1/y$. Generalize.

11. (a) Derive Eq. (2.14).

    (b) A more realistic model of the trout propagation problem would divide the fish population into (at least) two classes, small fish and large fish, with different mortality rates. Thus there would be two populations, $N_s(t)$ and $N_\ell(t)$. Try to develop a pair of (probably) coupled IFKs for these. Observe that both r and s contribute only to N_s. You will find that you must include some mechanism for allowing fish to grow from small to large.

12. Solve Eq. (2.22) for $H'(y)$. Notice that

$$\left(\frac{ds}{dy}\right)^2 = (H'(y))^2 = 1 + \left(\frac{dx}{dy}\right)^2.$$

Solve this differential equation for x as a function of y. You will probably not recognize the result as a cycloid because that curve is usually given parametrically. With the help of an analytic geometry book try to verify that you *do* have cycloid.

13. (a) Solve Eq. (2.22) where T is replaced by a specified function of y, $\tilde{T}(y)$.

    (b) In 13(a) let $\tilde{T}(y) = T + \epsilon(y)$. This corresponds to the experiment described at the end of Section 2.11, T being constant and $\epsilon(y)$ being small. Discuss.

14. The Chebyshev polynomials mentioned in Section 2.6 have the properties that $T_n(z)$ is a polynomial of degree n and

$$\int_{-1}^{1} \frac{T_n(z) T_m(z)}{\sqrt{1-z^2}} dz = 0, \qquad n \neq m.$$

Thus the T_n's form an orthogonal set on the interval $(-1, 1)$ with weight function $(1 - z^2)^{-1/2}$. Explicitly, $T_0(z) = 1$, $T_1(z) = z$, and for $n \geq 2$, $T_n(z) = 2zT_{n-1}(z) - T_{n-2}(z)$.

Another integral equation which arises in some tomographic studies is

$$(**) \qquad \tilde{g}(x) = \int_{-x}^{x} \frac{T_{2n}(\frac{y}{x})\tilde{f}(y)}{\sqrt{x^2 - y^2}} dy.$$

Show that if $\tilde{f}(y)$ solves this problem then so does $\tilde{f}(y) + Cy^k$ for any $k < 2n$ and any C. Thus Eq. (**) does not have a unique solution.

CHAPTER 3

A Bit of Functional Analysis

3.1 Introduction

In the previous chapter I alluded to "large" errors and functions. I also implied that numerical differentiation is often an unsatisfactory operation. This is a reflection of the common observation that differentiation is a "roughening" process, whereas integration is a "smoothing" one. For a better understanding of IFKs, it is desirable to develop a more satisfactory notion of the implied meaning of such terms as large, small, smooth, and rough.

3.2 Norms of Functions

Recall from our basic set of ground rules (Section 1.3) that all functions involved in this paper are nice—continuous, piecewise continuous, mildly unbounded, or the like. We first introduce the concept of the size, or more properly speaking, the concept of the *norm* of such a function.

Let us suppose that $h(x)$ is a bounded function for $a \leq x \leq b$. A measure of its size is simply the least upper bound, or supremum, of its absolute value in the interval $a \leq x \leq b$:

$$(3.1) \qquad \| h \|_S = \sup_{a \leq x \leq b} | h(x) | .$$

The number $\| h \|_S$ is defined as the sup-norm of h.

It is clear that the sup-norm does not exist as a finite number if $h(x)$ is not bounded. An example of this is

$$h_0(x) = \begin{cases} 0, & -1 \leq x \leq 0, \\ x^{-1/2}, & 0 < x \leq 1. \end{cases}$$

As x approaches zero from the right, $h_0(x)$ increases without bound. However, we note that

$$\int_{-1}^{1} |h_0(x)| \, dx = \int_0^1 \frac{dx}{\sqrt{x}} = 2.$$

This suggests the definition of a *new* norm

(3.2) $$\| h \|_1 = \int_a^b |h(x)| \, dx.$$

We refer to $\| h \|_1$ as the L_1-norm of h, or simply as the one-norm of h. Note that for the nice functions we are considering, any function with a sup-norm has a one-norm. (Recall that a and b are required to be finite.) The converse, clearly, is not true, as the example h_0 shows.

Another much used norm is given by

(3.3) $$\| h \|_2 = \left\{ \int_a^b |h(x)|^2 \, dx \right\}^{1/2}.$$

This is the L_2-norm, or simply the two-norm. Observe that $h_0(x)$ has a one-norm but *not* a two-norm.

There are many other possibilities. In fact, all that is required of a general norm $\| \cdots \|$ is that it be a real nonnegative number such that

(i) $\| h \| = 0$ if and only if $h \equiv 0$;

(ii) $\| ch \| = |c| \| h \|$ for c any constant;

(iii) $\| h_1 + h_2 \| \leq \| h_1 \| + \| h_2 \|$.

The square root occurs in Eq. (3.3) in order that properties (ii) and (iii) hold.

Which norm is best to use? The answer is problem dependent. In practice, we customarily select a *normed linear function space* in which to work. This is a set S of functions such that a real constant times a function in $\mathcal{S}$ is also a function in $\mathcal{S}$, the sum of any two functions in $\mathcal{S}$ is a function in $\mathcal{S}$, and all functions in $\mathcal{S}$ have a norm of the same kind. Thus we may deal with L_1-space, L_2-space, etc. Often the index (1, 2, etc.) is omitted from the norm notation when no confusion is likely to arise.

3.3 Norms of Vectors

The concept of a vector norm is really more familar than that of a function norm. The length (properly, the Euclidean length) of the vector

$$\mathbf{h} = h_1\mathbf{i} + h_2\mathbf{j} + h_3\mathbf{k}$$

is well known to be just

$$\sqrt{h_1^2 + h_2^2 + h_3^2}.$$

To generalize this idea to a vector in n-dimensional space (bold face is always used to indicate a vector),

$$\mathbf{h} = \begin{pmatrix} h_1 \\ h_2 \\ \vdots \\ h_n \end{pmatrix} = (h_1, h_2, \cdots, h_n)^T;$$

we write

$$\|\mathbf{h}\|_2 = \left\{ \sum_{j=1}^{n} |h_j|^2 \right\}^{1/2}.$$

This is an obvious analogue of the two-norm introduced in the previous section.

To extend the analogy further, we write

$$\|\mathbf{h}\|_S = \sup_{1 \leq j \leq n} |h_j|$$

and

$$\|\mathbf{h}\|_1 = \sum_{j=1}^{n} |h_j|.$$

Each of these norms can be shown to satisfy conditions (i)–(iii) of Section 3.2.

Note that all the common norms of a vector always exist, assuming, as we always shall, that all components are finite numbers, and the vector has a finite number of components.

3.4 Norms of Integral Operators

Let us turn to the integral operator that occurs in an IFK; see Eq. (1.2). For convenience we will suppose that $K(x,y)$ is a bounded function. (An easy extension of the concept introduced in Eq. (3.1) allows us to say then that K has a sup-norm.) Assuming also that we confine our investigations to functions f that possess a one-norm, we then define

$$(3.4) \qquad g(x) = \int_a^b K(x,y) f(y) dy.$$

The function g is meaningful for $a \leq x \leq b$. Ignore for the moment the fact that Eq. (3.4) can be viewed as an IFK. Instead, form

$$(3.5) \qquad \begin{aligned} \| g \|_1 &= \int_a^b | g(x) | \, dx = \int_a^b dx \left| \int_a^b K(x,y) f(y) dy \right| \\ &\leq \int_a^b dx \int_a^b | K(x,y) | \, | f(y) | \, dy \\ &\leq \| K \|_S \int_a^b dx \int_a^b | f(y) | \, dy = \| K \|_S (b-a) \| f \|_1. \end{aligned}$$

Therefore
$$\| g \|_1 \leq M_0 \| f \|_1.$$

Clearly, the function g has a one-norm. Although we have not found its value, we have calculated an upper bound as

$$(3.6) \qquad \| g \|_1 \leq M_0 \| f \|_1.$$

Thus the integration operator defined by Eq. (3.4) may change the one-norm of a function f but by no more than a certain amount, estimated by M_0. The integration operator is a *bounded* operator.

As an example, we select a special K:

$$K(x,y) = \begin{cases} 1, & a \leq y \leq x, \\ 0, & x < y \leq b. \end{cases}$$

A Bit of Functional Analysis

Equation (3.4) becomes

$$g(x) = \int_a^x f(y)dy.$$

Thus we are now talking about *ordinary* integration. The argument contained in the derivation of Eq. (3.5) leads immediately to

$$\|g\|_1 \leq (b-a)\|f\|_1.$$

Thus, g has a one-norm and we have estimated its size.

Finally, let us consider two different functions f_1 and f_2 that are close together. More accurately, we can now say

$$\|f_1 - f_2\|_1 - \epsilon,$$

where ϵ is small. Then, after calculating a little further, we find for the corresponding g's as defined by Eq. (3.4) that

$$\|g_1 - g_2\|_1 \leq M_0 \|f_1 - f_2\|_1 = \epsilon M_0.$$

This confirms our notion that integration is a smoothing operation. A small change in f results in a small change in g, but what intuition tells us is now expressed mathematically.

The number M_0 occurring in Eq. (3.6) may not be the smallest possible number. If we knew more about $K(x,y)$, we might be able to show that for *any* f with a one-norm

(3.7) $$\|g\|_1 \leq M_1 \|f\|_1$$

for some $M_1 < M_0$. If M_1 is the *smallest* possible number that may be used in Eq. (3.7) with all admissible f's, then we call M_1 the *one-norm* of the integral operator. Often one can demonstrate that a norm exists, but the determination of its actual value is usually quite difficult. Fortunately, in much of our work that value is not of great importance.

The use of the term norm applied to both functions or vectors and certain operators may be confusing initially. However, observe that the norm as newly defined is still a measure of size—it measures the size of the process of integration.

Although I have chosen to talk mainly about the one-norm in this section, most of my remarks apply to other norms where changes in the assumptions about $K(x,y)$ may be necessary. The two-norm theory is especially rich. In that case we need require only that

$$\int_a^b \int_a^b |K(x,y)|^2\, dx\, dy = M_2^2 < \infty$$

to conclude that the function g of Eq. (3.4) satisfies

$$\|g\|_2 \leq M_2 \|f\|_2,$$

provided f has a two-norm. We may even allow $a = -\infty$ and $b = +\infty$. I have confined most of this discussion to the one-norm because it is easy to handle technically and provides a clear idea about what is actually happening.

3.5 Norms of Differential Operators

Because differentiation is the inverse of integration, it is only reasonable to see if the ideas of the preceding section extend to this operation. Suppose that $f(x)$ is a differentiable function for $a \leq x \leq b$, and that it has a one-norm. Define

(3.8) $$g(x) = \frac{d}{dx} f(x).$$

Does g have a one-norm on $a \leq x \leq b$? Consider, as an example,

$$f(x) = \begin{cases} \log x, & 0 < x \leq 1, \\ 0, & x = 0. \end{cases}$$

Then f has a one-norm. It is given by

$$\|f\|_1 = \int_0^1 |\log x|\, dx = 1.$$

Now

$$g(x) = \frac{d}{dx} \log x = \frac{1}{x}, \qquad 0 < x \leq 1.$$

Because

$$\int_0^1 |g(x)|\, dx = \int_0^1 \frac{dx}{x}$$

does not exist, $\|g\|_1$ does not exist; neither does $\|g\|_S$ nor $\|g\|_2$, as may be easily demonstrated. The process of differentiation has yielded a function that possesses none of the norms we have been discussing.

A Bit of Functional Analysis

As another example, let
$$f_0(x) = x^{2/3}, \quad 0 \le x \le 1.$$
It is easily verified that f_0 has a sup-norm, a one-norm, and a two-norm. Write
$$g_0(x) = \frac{d}{dx} x^{2/3} = \frac{2}{3} x^{-1/3}, \quad 0 < x \le 1.$$
Clearly $\| g_0 \|_S$ does not exist. A bit of calculation shows, however, that $\| g_0 \|_1$ and $\| g_0 \|_2$ are perfectly good numbers.

Finally, if
$$f_1(x) = x^{1/4}, \quad 0 \le x \le 1,$$
then
$$g_1(x) = \frac{d}{dx} x^{1/4} = \frac{1}{4} x^{-3/4}, \quad 0 < x \le 1.$$
The function f_1 has all three of the norms under consideration, but g_1 has only a one-norm.

Evidently, the operation of differentiation is rather unpleasant. It can transform a nice function to one much less well behaved.

Now let us see if the operation of differentiation has a property analogous to that expressed by Eq. (3.6). Consider the sequence
$$f_n(x) = \sin n\pi x, \quad 0 \le x \le 1, \quad n = 1, 2, 3, \cdots$$
and the associated sequence of derivatives
$$g_n(x) = n\pi \cos n\pi x, \quad 0 \le x \le 1, \quad n = 1, 2, 3, \cdots$$
We find

(3.9) $$\| f_n \|_1 = \int_0^1 | f_n(x) | \, dx = \int_0^1 | \sin n\pi x | \, dx = \frac{2}{\pi},$$

and, for the corresponding derivatives,

(3.10) $$\| g_n \|_1 = \int_0^1 | g_n(x) | \, dx = n\pi \int_0^1 | \cos n\pi x | \, dx = 2n.$$

From Eqs. (3.9) and (3.10) we get

(3.11) $$\| g_n \|_1 = n\pi \, \| f_n \|_1 .$$

We conclude that there is *no* number M_0 for all n such that

$$\| g_n \|_1 \leq M_0 \| f_n \|_1 .$$

Thus, if we look at the differential operator acting on all those differentiable functions with one-norm and consider just those cases in which the derivative has a one-norm, Eq. (3.6) is not satisfied. Hence the derivative operator does not have a one-norm. (You will find it instructive to carry out similar investigations with other norms.)

Finally, let us suppose that h_1 and h_2 are close together in the one-norm sense,

$$\| h_1 - h_2 \|_1 = \epsilon.$$

Are the derivatives close together? To see that the answer may be "no," choose

$$h_1 - h_2 = \frac{\epsilon \pi}{2} \sin N\pi x, \qquad 0 \leq x \leq 1.$$

Then $\| h_1 - h_2 \|_1 = \epsilon$, but

$$\left\| \frac{d}{dx} h_1 - \frac{d}{dx} h_2 \right\|_1 = \pi \epsilon N.$$

By merely choosing N large, we see that the derivatives differ greatly in norm.

We have really mathematicized the intuitive notion that differentiation is a roughening operation. Of course, we might hope to find *some* norm (perhaps quite different from any of the ones we have been using) in which this is not the case. But we must also recall that we should not only talk of norms but also of *function spaces* (see the last paragraph of Section 3.2).

Suppose that we are able to confine our investigations to functions f that are linear, a highly restrictive function space. Thus

$$f(x) = ax + b, \qquad 0 \leq x \leq 1$$

and

$$g(x) = \frac{d}{dx} f(x) = a, \qquad 0 \leq x \leq 1.$$

For variety we consider the two-norm

$$\| f \|_2^2 = \int_0^1 (ax + b)^2 dx = \frac{a^2}{3} + ab + b^2 = \frac{a^2}{12} + \left(\frac{a}{2} + b \right)^2$$

and
$$\|g\|_2^2 = \int_0^1 a^2 dx = a^2.$$

But
$$a^2 \leq 12\left\{\frac{a^2}{12} + \left(\frac{a}{2}+b\right)^2\right\}.$$

Thus
$$\|g\|_2 \leq \sqrt{12}\,\|f\|_2,$$

so that the analogue of Eq. (3.6) is satisfied with $M_0 = \sqrt{12}$. Thus in this restricted space the derivative operator has a norm. The result agrees with intuition. Differentiation cannot roughen a linear function.

These remarks may be confusing at first reading. The existence of an operator norm is dependent upon the collection of functions we use as well as upon the norm we use. It is clear that the choice of norms and function spaces must be decided in advance. Suffice it to say that for most ordinary norms and for most reasonably large collections of functions the differentiation operator does not have a norm, but the integration operator does. The integration operator is thus usually *bounded*. The differentiation operator is usually *unbounded*. This observation is very important.

3.6 Norms of Matrix Operators

We turn to a few brief remarks about matrix operators. I consider only square matrices (of finite order n) with elements k_{ij}. I shall use boldface capital letters to denote such matrices. Consider, analogous to Eq. (3.4),

(3.12) $$\mathbf{g} = \mathbf{Kf},$$

where $\mathbf{f}$ and $\mathbf{g}$ are n-vectors. Explicitly,

(3.13) $$g_i = \sum_{j=1}^n k_{ij} f_j, \quad i = 1, 2, \cdots, n.$$

Using the concept of the one-norm for vectors, we find that because

$$|g_i| \leq \sum_{j=1}^n |k_{ij}|\,|f_j| \leq C_i \sum_{j=1}^n |f_j| = C_i\,\|\mathbf{f}\|_1,$$

where

$$C_i = \sup_{1 \leq j \leq n} |k_{ij}|,$$

we have

$$\| \mathbf{g} \|_1 = \sum_{i=1}^{n} |g_i| \leq \sum_{i=1}^{n} C_i \| \mathbf{f} \|_1 = \| \mathbf{f} \|_1 \sum_{i=1}^{n} C_i = M_0 \| \mathbf{f} \|_1.$$

Note that M_0 is independent of $\mathbf{f}$. Just as in the case of integral operators, it may be possible to choose a number $M_1 < M_0$ such that

(3.14) $$\| \mathbf{g} \|_1 \leq M_1 \| \mathbf{f} \|_1.$$

If M_1 is the *smallest* possible value for which Eq. (3.14) holds for *all* vectors $\mathbf{f}$, then M_1 is called the one-norm of the matrix operator $\mathbf{K}$.

It is interesting to observe that we have obtained the existence of the norm of $\mathbf{K}$ without any special restrictions on $\mathbf{K}$. (Contrast this with the situation in Section 3.4 where we did place conditions on $K(x,y)$.) Indeed, $\mathbf{K}$ also has a sup-norm and a two-norm (and just about any other norm we wish to deal with!). All of this is true basically because $\mathbf{K}$ is a finite-dimensional operator, although I do not choose to pursue the details. The matrix operator $\mathbf{K}$ is a *bounded* operator.

Now suppose that $\mathbf{K}$ has an inverse. That inverse will also be a matrix of the same type. It then will have a norm of almost any kind we wish. Compare this with the vastly different integration-differentiation situation. We shall see later that this sharp contrast is basic to the difficulties experienced in trying to handle IFKs numerically.

3.7 Norms of General Operators T

Although we shall not involve ourselves too deeply with operators different from those already discussed, it may be worthwhile to introduce the general concept of an operator norm. Let us consider a quite general operator T, a space $\mathcal{S}$ of functions f, and a particular norm $\| \cdots \|$. Let

$$g = Tf$$

and suppose that g lies in the same space $\mathcal{S}$ as f (roughly speaking, they are the same "kinds" of functions). Thus we may form $\| g \|$ as well as

$\| f \|$. Suppose further that there is a number M_0 such that for all f in S

$$\| g \| = \| Tf \| \leq M_0 \| f \|.$$

Then T is called a *bounded* operator. If M_1 is the smallest number such that for all such f

$$\| Tf \| \leq M_1 \| f \|,$$

then M_1 is called the *norm* of T and we write $M_1 = \| T \|$. If T is not bounded, then it is called *unbounded*.

Putting the results of the last several sections in these new terms, and speaking quite imprecisely, matrix operators are bounded, integration operators are (usually) bounded, and differential operators are (usually) unbounded.

I call attention to the fact that no specific representation or way of writing the operator T has been introduced. The boundedness or unboundedness of an operator is totally independent of such representation. Returning to the inverse Laplace transform (Section 2.2), we recall the two representations (2.2a) and (2.2b). The first *looks* perfectly nice, whereas the second reveals all sorts of difficulties. The inverse Laplace transform operator inherently contains all difficulties that are so apparent in the one representation and so neatly concealed in the other. In any reasonable space and norm this inverse Laplace transform operator is *unbounded*.

I also emphasize that the existence or nonexistence of a norm has nothing to do with numerics. No numerical approximations have been used in our discussions. We shall see as we go on that numerical methods sometimes lead to bounded approximate operators although the original operator is unbounded. This is a source of very considerable difficulty.

3.8 Summary

I have presented in this chapter a rather casual discussion of ways in which the size of functions and operators can be measured. These concepts will prove of great value in the work that follows. Of special importance will be the observation that certain operators of interest to us are unbounded. It is this fact that makes it so unpleasant to work with the IFK.

If you are interested in learning more about norms, bounded operators, etc., you should consult any book on functional analysis. (See, e.g., [2, 9, 32, 43, 49].)

Problems III

1. Find, or at least determine an upper bound on the sup-norm, one-norm, and two-norm of each of the following functions when these norms exist.

   (a) $h(x) = \frac{\sin x}{x}, \qquad -\pi \leq x \leq \pi;$

   (b) $h(x) = \frac{\cos x}{\sqrt{x}}, \qquad 0 < x \leq 2\pi;$

   (c) $h(x) = \frac{x}{(1-x^2)^4}, \qquad 0 \leq x \leq \frac{1}{2};$

   (d) $h(x) = e^{-1/x}, \qquad 0 < x \leq 1.$

2. (This problem represents a temporary suspension of our ground rules.) Suppose that $h(x)$ has a norm (sup-, one-, or two-) on the interval $(0, X)$ for arbitrarily large X. Does it have a norm of the same kind on $(0, \infty)$? Justify your answers.

3. Assume that h has a sup-norm on $(0, 1)$. Consider the function $q(x) = \int_0^1 K(x,y)h(y)dy$. For each of the following Ks determine if $q(x)$ has a sup-norm, a one-norm, or a two-norm on $0 \leq x \leq 1$.

   (a) $K(x, y) = \sin(xy), \qquad 0 \leq x, y \leq 1;$

   (b) $K(x, y) = e^{-|x-y|}, \qquad 0 \leq x, y \leq 1;$

   (c) $K(x, y) = \begin{cases} \frac{1}{\sqrt{x-y}}, & 0 < y < x, \\ 0, & x \leq y \leq 1. \end{cases}$

   Are you prepared to say in any instance that you have actually found the operator norm?

4. Repeat Problem 3 under the assumption that $h(x)$ has a one-norm on $(0, 1)$. (Note that in some of these problems we are "mixing" norms. Sometimes this is undesirable.)

5. An important inequality in analysis is due to Schwarz. It states that if $h_1(x)$ and $h_2(x)$ have two-norms on $(0,1)$ then $h_1(x)h_2(x)$ has a one-norm and

$$\int_0^1 | h_1(x)h_2(x) | \, dx \leq \left\{ \int_0^1 h_1^2(x) dx \right\}^{1/2} \left\{ \int_0^1 h_2^2(x) dx \right\}^{1/2}$$

(see [28]). Repeat Problem 4 using two-norms in all cases.

6. Check on the existence or nonexistence of all norms discussed in Section 3.5.

7. Study the question of the existence of the sup-norm and two-norm of the differential operator.

8. What is the vector analogue of Schwarz's inequality (Problem 5)? Try to understand this geometrically in the case that **h** has just two components. Discuss the two-norm of the matrix operator **K**.

9. In Section 3.7 there is a brief discussion of "general" operators. For example, we might have

$$Tf = \frac{d}{dx} f(x) + \int_a^b K(x,y) f(y) dy.$$

Try to think of other "general" operators you may have encountered in practice.

10. Show that the following sets S form normed linear function spaces for the norm indicated.

    (a) S the set of all polynomials on $0 \leq x \leq 1$, one-norm.
    (b) S the set of all trigonometric series with a finite number of terms on $-\pi \leq x \leq \pi$, sup-norm
    (c) The same as 10(b) but with two-norm.

11. Consider functions $h(x)$ defined on $(0,1)$ and with a one norm. Are the following valid norms?

    (a) $\| h \| = \int_0^{1/2} | h(x) | \, dx$;
    (b) $\| h \| = 17 \int_0^1 | h(x) | \, dx$;
    (c) $\| h \| = \int_0^1 \sin 6\pi x \, | h(x) | \, dx$.

CHAPTER 4

Integral Operators with Separable Kernels

4.1 Introduction

In this chapter I discuss a special kind of integral operator and some equations to which it leads. There are at least two justifications for this specialization. First, the kernels involved sometimes occur directly in applications. Also, they often can be used as approximations to more complicated kernels. Second, the theory for these operators reduces in a natural way to the theory for matrix operators, which are relatively easy to analyze. This study leads then to conjectures concerning more general integral operators. As we shall see in the next chapter, some of these conjectures turn out to be correct and others are quite false. It is those false ones that are interesting and frustrating. If the whole theory of IFKs could be reduced to matrix analysis, life would be much easier.

4.2 Separable Kernels

We confine our study in this chapter to kernels of the form

$$(4.1) \qquad K(x,y) = \sum_{j=1}^{n} \alpha_j(x)\beta_j(y).$$

Such a kernel is called *separable*. (The term *degenerate* is also used.) Observe that the sum involves only a finite number of terms. Moreover, we may assume that the set of functions $\alpha_j(x)$ is a linearly independent set. Similarly, the set of β_j is such a set. If this is not the case, the sum may be rewritten in such a way as to make it so and the α's and β's are then redefined. We could also orthonormalize each of these sets. This will not be done because certain of the ideas involved in what follows would be slightly obscured.

Finally, we assume that the functions $\alpha_j(x)$ are all nice for $a \leq x \leq b$ and that the β's are similarly well behaved and defined for $a \leq y \leq b$. Thus $K(x,y)$ is defined on the square $a \leq x, y \leq b$. We have actually made a similar assumption earlier concerning K (see Sections 1.1 and 1.4).

Before proceeding, we give two examples of separable kernels:

1. $K(x,y) = 1 + xy$,

2. $K(x,y) = a_0/2 + \sum_{j=1}^{n} \{a_j \cos[j(x-y)] + b_j \sin[j(x-y)]\}$, a_j and b_j given.

You will recognize (2) as a rather special truncated Fourier series. This, in turn, might be an approximation to a more complicated kernel. We shall eventually find some interesting facts concerning such approximations.

4.3 Some Properties of Integral Operators with Separable Kernels

For the moment we consider f to be a *given* function and define

$$g(x) = \int_a^b K(x,y) f(y) dy = \sum_{j=1}^{n} \int_a^b \alpha_j(x) \beta_j(y) f(y) dy$$

(4.2)
$$= \sum_{j=1}^{n} \alpha_j(x) \int_a^b \beta_j(y) f(y) dy = \sum_{j=1}^{n} c_j \alpha_j(x),$$

where

$$c_j = \int_a^b \beta_j(y) f(y) dy.$$

Equation (4.2) shows at once that the functions g that can be represented by this integral operator *must* be of a particular form; namely, g must be a linear combination of the α's. This is a highly restricted function space of g's.

Next, we note that there are many functions ψ that are orthogonal to the set of β's,

$$\int_a^b \beta_j(y)\psi(y)dy = 0, \qquad j = 1, 2, \cdots, n.$$

This is a reflection of the fact that the β set is finite. Define $f_1 = f + \psi$. Then

$$\begin{aligned}
\int_a^b K(x,y)f_1(y)dy &= \sum_{j=1}^n \int_a^b \alpha_j(x)\beta_j(y)\{f(y) + \psi(y)\}\,dy \\
&= \sum_{j=1}^n \int_a^b \alpha_j(x)\beta_j(y)f(y)dy \\
&\quad + \sum_{j=1}^n \alpha_j(x) \int_a^b \beta_j(y)\psi(y)dy \\
&= \sum_{j=1}^n \int_a^b \alpha_j(x)\beta_j(y)f(y)dy = g(x).
\end{aligned}$$

Thus there exist many functions f_1 that give rise, through the integral operator, to the same function g. Uniqueness is totally lacking.

You may be wondering why we do not simply move to a study of kernels representable by *infinite* sums. There are several reasons. For example, mathematical niceties of convergence properties and the like arise, and in a very nontrivial way. Perhaps more important is the fact that we wish to employ an understanding of finite matrices. Infinite sums would lead to matrices of infinite order, and these have many of the unpleasant properties of the general integral operators we shall be investigating.

4.4   IFKs with Separable Kernels

Now let us reconsider Eq. (4.2), this time supposing g to be known while f is to be determined. The remarks of the previous section lead at once to two observations:

1. g must be a linear combination of the α's or the problem is insoluble, and

2. If the problem is soluble, it has infinitely many solutions.

Although these facts now seem obvious, they are often overlooked.

Now assume that g is of the proper form (or, equivalently, that it belongs to the proper space),

(4.3) $$g(x) = \sum_{j=1}^{n} g_j \alpha_j(x).$$

Can we find an f? We *guess*

(4.4) $$f(y) = \sum_{i=1}^{n} f_i \beta_i(y),$$

where the f_i are constants to be determined, if possible. We next compute (see Eq. (4.2))

$$\begin{aligned} g(x) &= \sum_{j=1}^{n} g_j \alpha_j(x) = \sum_{j=1}^{n} \alpha_j(x) \int_a^b \left\{ \sum_{i=1}^{n} f_i \beta_i(y) \beta_j(y) \right\} dy \\ &= \sum_{j=1}^{n} \alpha_j(x) \sum_{i=1}^{n} f_i \int_a^b \beta_i(y) \beta_j(y) dy = \sum_{j=1}^{n} \alpha_j(x) \sum_{i=1}^{n} b_{ji} f_i, \end{aligned}$$

where

$$b_{ji} = \int_a^b \beta_i(y) \beta_j(y) dy.$$

Since the α's have been chosen to form a linearly independent set and

$$\sum_{j=1}^{n} \left\{ g_j - \sum_{i=1}^{n} b_{ji} f_i \right\} \alpha_j(x) = 0,$$

it follows that

$$g_j = \sum_{i=1}^{n} b_{ji} f_i$$

or, in matrix form,

$$\mathbf{g} = \mathbf{Bf},$$

where

$$\mathbf{B} = (b_{ji}).$$

It may be shown that the matrix $\mathbf{B}$ is nonsingular. ($\mathbf{B}$ is the Grammian matrix associated with the set β_j; see [11].) Thus the vector $\mathbf{f}$ is uniquely determined, and we have a unique f of the form Eq. (4.4). This is *not* the unique solution of the IFK. As noted earlier, the IFK does not have a unique solution. We *have* established the fact that it has *a* solution for any g of the proper form.

4.5 Eigenvalues and Eigenfunctions of Integral Operators with Separable Kernels

This and the following section may seem to be digressions. Actually they will provide interesting and valuable results. That we now turn to a discussion of eigenvalues should not really be too surprising. The preceding problem has been reduced to one in matrix theory, and eigenvalue and eigenvector considerations are of prime importance in that theory. It is helpful to study their analogues for integral operators.

We consider the equation

$$\lambda \phi(x) = \int_a^b K(x,y)\phi(y)dy, \tag{4.5}$$

where K is still given by Eq. (4.1). In view of the observation in Section 4.3, any function ϕ satisfying Eq. (4.5) must have the form

$$\lambda \phi(x) = \lambda \sum_{j=1}^n \phi_j \alpha_j(x). \tag{4.6}$$

Again we compute:

$$\lambda \phi(x) = \lambda \sum_{j=1}^n \phi_j \alpha_j(x) = \sum_{j=1}^n \alpha_j(x) \int_a^b \beta_j(y) \sum_{p=1}^n \phi_p \alpha_p(y) dy$$

$$= \sum_{j=1}^n \alpha_j(x) \sum_{p=1}^n \gamma_{jp} \phi_p,$$

with

$$\gamma_{jp} = \int_a^b \alpha_p(y) \beta_j(y) dy.$$

Using arguments identical to those of the previous section we find

$$\lambda \phi = \Gamma \phi, \tag{4.7}$$

where

$$\Gamma = (\gamma_{jp}).$$

The matrix system Eq. (4.7) has n eigenvalues and n eigenvectors. Their exact properties may be discussed in terms of the structure of Γ, which in turn inherits its structure from K. In particular, if it should happen that $\alpha_j(x) = \beta_j(x)$, $j = 1, 2, \cdots, n$, then $K(x,y) = K(y,x)$.

Thus K is a *symmetric* kernel and Γ is a symmetric matrix. In this case the eigenvalues are all real, the eigenvectors are linearly independent, etc. Each of these properties translates into a corresponding property for the IFK through the relationship Eq. (4.6).

It must be noted that this procedure conceals many eigenfunctions of the original problem. For instance, suppose ψ is *any* function orthogonal to all the β's. Then

$$\int_a^b K(x,y)\psi(y)dy = \sum_{j=1}^n \alpha_j(x) \int_a^b \beta_j(y)\psi(y)dy = 0 = 0 \cdot \psi(x),$$

and so such a ψ is an eigenfunction with eigenvalue zero. (A review of Eq. (4.6) and the subsequent calculations reveals no inconsistency.) However, our approach produces all eigenfunctions not having eigenvalue zero.

I prefer not to pursue this matter further. At this point you may find it desirable to invent a rather simple separable kernel and study its properties in detail by paralleling the theory (see, e.g., Problem 5). Actually, a great deal more may be learned about integral operators by studying properties of the matrices in the separable case. Some of these facts about matrices are quite possibly unfamiliar to you. Because our primary objective is the understanding of IFKs, I prefer to introduce such unfamiliar matters in the context of integral operators rather than in the matrix context.

4.6 Integral Equations of the Second Kind with Separable Kernels

We turn now to a brief investigation of the equation of the second kind (see Eq. (1.1)),

(4.8) $\qquad f(x) = g(x) + \gamma \int_a^b K(x,y)f(y)dy, \qquad \gamma \neq 0,$

where K remains defined by Eq. (4.1). A trivial rewriting of Eq. (4.8) yields

$$f(x) - g(x) = \gamma \sum_{j=1}^n \alpha_j(x) \int_a^b \beta_j(y)f(y)dy = \sum_{j=1}^n h_j \alpha_j(x),$$

(4.9) $\qquad h_j = \gamma \int_a^b \beta_j(y)f(y)dy.$

Thus we must have

$$(4.10) \qquad f(x) = g(x) + \sum_{p=1}^{n} h_p \alpha_p(x).$$

We proceed to calculate the unknown coefficients h_p:

$$g(x) + \sum_{j=1}^{n} h_j \alpha_j(x)$$

$$= g(x) + \gamma \sum_{j=1}^{n} \alpha_j(x) \int_a^b \beta_j(y) \left[g(y) + \sum_{p=1}^{n} h_p \alpha_p(y) \right] dy$$

$$= g(x) + \sum_{j=1}^{n} \alpha_j(x) \left[g_j + \gamma \sum_{p=1}^{n} h_p \gamma_{jp} \right],$$

where

$$g_j = \gamma \int_a^b g(y) \beta_j(y) dy, \qquad \gamma_{jp} = \int_a^b \alpha_p(y) \beta_j(y) dy.$$

Thus

$$\sum_{j=1}^{n} h_j \alpha_j(x) = \sum_{j=1}^{n} \alpha_j(x) \left[g_j + \gamma \sum_{p=1}^{n} \gamma_{jp} h_p \right],$$

so that for $j = 1, 2, \cdots, n$,

$$h_j = g_j + \gamma \sum_{p=1}^{n} \gamma_{jp} h_p.$$

In matrix form
$$(4.11) \qquad \mathbf{h} = \mathbf{g} + \gamma \Gamma \mathbf{h},$$

where Γ is the same matrix as in Eq. (4.7). Equation (4.11) can be rewritten

$$(4.12) \qquad (\mathbf{I} - \gamma \Gamma)\mathbf{h} = \mathbf{g},$$

where $\mathbf{I}$ is the identity matrix. Equation (4.12) can be solved *uniquely* for the unknown coefficients h_p in Eq. (4.10) provided γ is not the reciprocal of an eigenvalue of the matrix Γ. This provides the *unique* solution f of Eq. (4.8). It should be remarked that under certain conditions (which I shall not detail) Eq. (4.12) can be solved even if γ is the reciprocal of an eigenvalue of Γ. The resulting f is *not* unique.

4.7 Summary

In this chapter we have studied integral operators with kernels of a particularly simple, yet useful, type. A close connection has been found between these operators and matrix operators. That connection has been exploited to solve not only the IFK associated with such integral operators, but also the associated eigenvalue-eigenfunction problem and integral equations of the second kind. A considerable lack of uniqueness has been discovered. This lack of uniqueness is least distressing in the solution of integral equations of the second kind because it can appear only when the parameter γ happens to have one of a finite number of specific values.

In the next chapter we look at integral operators defined by much more general kernels and investigate similar problems.

Problems IV

1. Write the kernel
$$K(x,y) = x^2 + (x+y+2)^2 - xy$$
in the form Eq. (4.1) with the functions α_j and β_j linearly independent.

2. Does the equation
$$x^2 + 1 = \int_0^1 (xy - 7) f(y) dy, \qquad 0 \le x \le 1,$$
have a solution? Explain.

3. Let
   (a) $\beta_j(y) = \sin jy$, $j = 1, 2, \cdots, 10$, $\qquad 0 \le y \le \pi$;
   (b) $\beta_j(y) = y^j$, $j = 0, 1, \cdots, 100$, $\qquad -1 \le y \le 1$.

   In each case find several functions $\psi(y)$ orthogonal to all the β's.

4. Find a solution to each of the following equations, using the approach of Section 4.4.

   (a) $1 = \int_0^1 (xy + 17) f(y) dy$, $\qquad 0 \le x \le 1$;

(b) $\sin x + 3\sin 2x + \cos 2x = \int_0^\pi [\sin x \cos y + \sin 2(x+y)] f(y)dy$, $0 \le x \le \pi$.

5. Find the nonzero eigenvalues and corresponding eigenfunctions of each of the following kernels.

   (a) $K(x,y) = xy + 17$, $\quad 0 \le x, y \le 1$;
   (b) $K(x,y) = \sin 2(x+y)$, $\quad -\pi \le x, y \le \pi$.

6. Solve completely, using the method of Section 4.6.

   (a) $f(x) = x + \gamma \int_0^1 (xy + 17) f(y) dy$, $\quad 0 \le x \le 1$;
   (b) $f(x) = 1 + \gamma \int_{-\pi}^{\pi} [\sin 2(x+y)] f(y) dy$, $\quad -\pi \le x \le \pi$.

   Are there any exceptional values of γ? Discuss uniqueness matters in any such cases.

7. Consider the kernel
$$K_n(x,y) = \sum_{j=1}^n \frac{x^j y^j}{2^j j}, \quad 0 \le x, y \le 1.$$

   Observe that the elements γ_{jp} of the matrix Γ (see Eq. (4.7)) can be computed analytically. Do so. Then with the aid of a computer solve the corresponding eigenvalue-eigenfunction problem for various values of n. Are you prepared to make any conjectures? If your computer has the capability, plot some of the eigenfunctions. Compare eigenfunctions of the same index using larger and larger n.

   You have probably noticed that for large n, K_n and K_{n+1} are almost equal. Moreover, for $0 \le x, y \le 1$
$$\lim_{n \to \infty} K_n(x,y) = -\log\left(1 - \frac{xy}{2}\right).$$
   Any further conjectures?

8. Using K_n as in Problem 7 study the solution of
$$f(x) = e^x + \int_0^1 K_n(x,y) f(y) dy,$$
   using a numerical quadrature scheme to compute the elements g_j. Again vary n.

9. (a) Apply the method of Section 4.4 to obtain a solution to

$$x = \int_0^1 K_n(x,y) f(y) dy,$$

K_n as above, for various values of n. Notice that this equation has a nonunique solution.

(b) Suppose the left-hand side of Problem 9(a) is contaminated with error. Simulate this by adding a small random number R_j to each g_j. Notice that the integral equation no longer has a solution but the matrix equation of Section 4.4 can still be solved. Carry out the numerical calculations for errors of various magnitudes and different values of n. Discuss. (Note: If your computer software does not have a random number generator, try $R_j = \epsilon \sin[(N+j)!]$ using a different integer N for each experiment.)

CHAPTER 5

Integral Operators with General Kernels

5.1 Introduction

In this chapter we investigate operators with quite general kernels, always with the restriction that those kernels are fairly well-behaved. Recall that some properties of integral operators have already been discovered in Chapter 3. In particular, we have found that such operators are ordinarily bounded. We shall study other properties such as invertibility and boundedness or unboundedness of the inverse when it exists.

In contrast to the previous chapter, I delay discussion of IFKs until the end, taking up eigenvalue questions and even some discussion of integral equations of the second kind before turning to the topic of major interest. In much of the chapter I merely state results, sometimes providing relatively crude and intuitive justifications. References will be provided for those who wish a more complete treatment. The study of separable kernels will be helpful, although sometimes a bit misleading.

It should be noted that separable kernels form a (small) subset of the kernels under discussion here. All of the observations in this chapter therefore apply to such kernels.

5.2 Some More Facts About Integral Operators—Existence and Boundedness of Inverses

A somewhat improved notation will now be useful. I shall often henceforth write

(5.1a) $$Kf = \int_a^b K(x,y) f(y) dy,$$

or, even more symbolically,

(5.1b) $$K \cdot = \int_a^b K(x,y) \cdot dy.$$

You should avoid the temptation to identify K with a matrix. The latter will always be distinguished by a letter in boldface.

The introduction of this notation leads me to note explicitly that throughout this primer we have been studying *linear operators*. I have used, without specific mention, the relation

$$\int_a^b K(x,y) \{C_1 f_1(y) + C_2 f_2(y)\} dy = C_1 \int_a^b K(x,y) f_1(y) dy + C_2 \int_a^b K(x,y) f_2(y) dy,$$

where C_1 and C_2 are arbitrary constants. This now becomes

(5.2) $$K(C_1 f_1 + C_2 f_2) = C_1 K f_1 + C_2 K f_2.$$

Equation (5.2) is the mathematical statement that K is a *linear operator*. (So is the matrix operator $\mathbf{K}$.) From here on, I shall often employ Eq. (5.2) without comment. Thus linearity will be everywhere assumed.

We next investigate the question of the inverse operator. Is there an operator, which I denote by K^{-1}, such that if

$$g = Kf$$

then

$$f = K^{-1}g?$$

We shall require that this inverse be unique; only *one* g can be such that $f = K^{-1}g$.

The linearity leads to an important observation. Namely, if K^{-1} exists, there is no $h \neq 0$ such that

(5.3) $$Kh = 0,{}^1$$

for linearity implies
$$K0 = 0.$$

If Eq. (5.3) holds for $h \neq 0$ our uniqueness requirement is violated.

An example of interest is the kernel

$$K(x,y) = \cos(xy), \quad -1 \leq x, \quad y \leq 1.$$

Let $f(y)$ be any *odd* function, $f(y) = -f(-y)$. Because the cosine is an even function, we have

$$Kf = \int_{-1}^{1} \cos(xy) f(y) dy = 0.$$

Thus K cannot have an inverse.

At this point, I introduce a note of caution. Suppose we place a restriction on the functions to which we apply the cosine kernel and require that only *even* functions will be admissible. Thus we confine the functions f to a certain space. It is not clear whether

$$Kf = \int_{-1}^{1} \cos(xy) f(y) dy = 0, \quad -1 \leq x \leq 1,$$

for some such even $f \neq 0$. Suppose there is no such even f. Can we *then* assert that K^{-1} exists?

To answer this question we must examine the equation

$$\int_{-1}^{1} \cos(xy) f(y) dy = g(x)$$

and ask if there is an even f that produces the function y, whatever g may be. The answer is "no." For $\cos(xy)$ is an even function of x. Therefore g must be even.

Let us now restrict *both* f and g to be even. Does K have an inverse? We still cannot say without a much deeper investigation which would

---

[1] I write $h \neq 0$ to indicate $h(x) \not\equiv 0$. Similarly $h = 0$ if $h(x) \equiv 0$. An analogous notation will be used for vectors and matrices.

take us too far afield. There may be a "reasonable" function g such that there is no "reasonable" function f for which

$$Kf = g.$$

The word *reasonable* suggests that we really should talk about the spaces to which the functions belong. That is precisely the case. As we have discussed earlier, the choice of spaces is always important.

To bring this matter into sharper focus, let us recall that if K is separable there is always a function $f \neq 0$ such that $Kf = 0$. Thus all the operators considered in the previous chapter appear to lack inverses. This seems contradictory because the matrix **B** associated with K in Section 4.4 is nonsingular and so it is invertible. The difficulty is resolved when we recall that **B** emerged after f and g were suitably restricted. Thus, appropriate spaces were chosen.

You may now be very confused and worried. What general statement can we make about inverses? Suppose we consider spaces $\mathcal{S}_1$ and $\mathcal{S}_2$ of functions f and g such that for f in $\mathcal{S}_1$ there is a g in $\mathcal{S}_2$ for which $Kf = g$. Suppose the converse is also true. Finally, suppose further that there is no f in $\mathcal{S}_1$, $f \neq 0$, such that $Kf = 0$. Then K has an inverse so long as we restrict our investigation to $\mathcal{S}_1$ and $\mathcal{S}_2$. The "proof" of this is essentially contained in the foregoing discussion.

Now, under the assumption that K has an inverse, let us try to decide whether that inverse operator is bounded or unbounded. Again, the result is strongly "space dependent." We turn once more to the previous chapter and consider the separable kernel case. We know that if g is not restricted, the inverse operator does not even exist. But when f and g belong to appropriate spaces, K is represented by the nonsingular matrix **B**, **B** is invertible, and the inverse matrix is, of course, a bounded operator (see Section 3.6). Hence K^{-1} is a bounded operator in these spaces.

From here on, unless specific mention is made of particular spaces, we shall ordinarily assume that f and g are simply "natural" functions (continuous, piecewise continuous, etc). or "natural" vectors (of appropriate dimensions). Thus f and g belong to their "natural" spaces.

We examine a more general problem. Suppose that $K(x, y)$ and $\partial K / \partial y$ both have sup-norms on $0 \leq x, y \leq 2\pi$. Consider

(5.4) $$g_n(x) = \int_0^{2\pi} K(x, y) \psi_n(y) dy,$$

$$\psi_n(y) = \cos ny.$$

It is easy to see that g_n has a sup-norm on $0 \leq x \leq 2\pi$. Let us assume K has a valid inverse and write

(5.5) $$\cos ny = K^{-1}g_n(x).$$

We ask if K^{-1} is a bounded operator in the space of functions with sup-norm.

First, notice that $\lim_{n\to\infty} g_n(x) = 0$. This follows from the fact that for *fixed* x, g_n is just the coefficient of the Fourier cosine series for K, considered as a function of y. (This is the so-called Riemann–Lebesgue theorem. See [32].) In fact, we can do better. Integrate Eq. (5.4) by parts:

$$\begin{aligned} g_n(x) &= K(x,y)\frac{\sin ny}{n}\Big|_{y=0}^{y=2\pi} - \frac{1}{n}\int_0^{2\pi}\frac{\partial K}{\partial y}(x,y)\sin ny\, dy \\ &= -\frac{1}{n}\int_0^{2\pi}\frac{\partial K}{\partial y}\sin ny\, dy. \end{aligned}$$

Thus
$$|g_n(x)| \leq \frac{2\pi}{n}\left\|\frac{\partial K}{\partial y}\right\|_S,$$

and so

(5.6) $$\|g_n\|_S \leq \frac{2\pi}{n}\left\|\frac{\partial K}{\partial y}\right\|_S.$$

Now *suppose* K^{-1} is a bounded operator. That is, $\|K^{-1}g_n\|_S \leq M_S\|g_n\|_S$. From Eqs. (5.5) and (5.6)

$$\|\cos ny\|_S = \|K^{-1}g_n(x)\|_S \leq M_S\|g_n(x)\|_S \leq \frac{2\pi M_S}{n}\left\|\frac{\partial K}{\partial y}\right\|_S$$

or

$$n\|\cos ny\|_S \leq 2\pi M_S\left\|\frac{\partial K}{\partial y}\right\|_S = \text{constant}.$$

However,
$$\|\cos ny\|_S = 1,$$

and we have an obvious contradiction for n large. Therefore K^{-1} *cannot be bounded* on the space of functions with sup-norms.

This result seems to have been obtained through a very careful choice of norms, kernel, etc. It can be greatly generalized (see Problems 3–5).

As suggested by Chapter 3, the inverse of an integral operator is ordinarily *not* bounded. The behavior of those inverses that were found in very special cases in Chapter 2 to contain derivatives (which are unbounded operators) is really quite general.

The above can be put into somewhat more homely and perhaps more instructive terms. Suppose

(5.7) $$g(x) = \int_a^b K(x,y)f(y)dy.$$

Then it is possible to add to f a "large" function (in our specific example $C\cos(ny)$, C arbitrary) and change g by an arbitrarily "small" amount. In the terminology of Section 2.12, it appears that quite general IFKs are ill posed. When we recall that (5.7) is precisely our IFK and that g is given at relatively few points, and there with inaccuracies, we can see that the troubles encountered in some of the examples of Chapter 2 are ubiquitous.

I shall try to summarize the foregoing, first noting that for a thorough understanding and exposition of the matters discussed we would have to delve much deeper into somewhat sophisticated mathematics. I prefer to proceed on an "intuitive" level. We have found that an integral operator does not have an inverse unless suitable restrictions are placed on the function spaces under consideration. A *necessary*, but by no means sufficient, condition for the existence of an inverse is that $K\psi = 0$ only if $\psi = 0$. If the inverse K^{-1} exists, it is ordinarily an unbounded operator unless restrictions are placed on the function spaces being used. Such restrictions must usually be quite severe.

5.3 Eigenvalues and Eigenfunctions of Integral Operators with Symmetric Kernels

In the case where $K(x,y)$ is a symmetric function, $K(x,y) = K(y,x)$, the operator K has some especially attractive properties. Such an operator is itself called *symmetric*. You will recall that symmetric matrices are much easier to handle than more general matrices; the same is true of symmetric integral operators. (Recall that we are dealing in this primer with real $K(x,y)$. For some remarks pertaining to the complex case, see Appendix B.)

We first consider the equation

(5.8) $$\lambda\phi(x) = \int_a^b K(x,y)\phi(y)dy.$$

The following facts are known (see, for example, [11, 33, 46]) in case K is symmetric and $K(x,y)$ is well-behaved. For example, if the condition $\int_a^b \int_a^b K^2(x,y)dxdy < \infty$ encountered briefly in Section 3.4 holds, then so do the following:

1. Equation (5.8) has at least one solution $\phi \neq 0$ and this corresponds to a value of $\lambda \neq 0$. The function ϕ and the constant λ are an *eigenfunction* and a corresponding *eigenvalue* of K.

2. If ϕ_1 and ϕ_2 belong to different eigenvalues λ_1 and λ_2, then ϕ_1 and ϕ_2 are orthogonal:
$$\int_a^b \phi_1(x)\phi_2(x)dx = (\psi_1,\psi_2) = 0.\text{[2]}$$

3. Two different eigenfunctions may belong to the same eigenvalue. However, a particular nonzero eigenvalue can have associated with it only a finite number of linearly independent eigenfunctions. These may be orthogonalized (and, of course, normalized).

4. Equation (5.8) has only a finite number of nonzero eigenvalues if and only if $K(x,y)$ is separable.

5. All eigenvalues are real and all eigenfunctions may be chosen to be real.

6. If there are infinitely many nonzero eigenvalues λ_n, then $\lim_{n\to\infty} \lambda_n = 0$. (It is customary to order the eigenvalues so that $|\lambda_{n+1}| \leq |\lambda_n|$.)

The existence of eigenfunctions immediately suggests that we may be able to expand an "arbitrary" functions in terms of these eigenfunctions:

(5.9) $$g(x) = \sum_{n=1}^{\infty} a_n \phi_n(x), \quad a_n = \int_a^b g(x)\phi_n(x)dx = (g,\phi_n).$$

---

[2] We assume that the reader is familiar with the inner product notation, $(h_1,h_2) = \int_a^b h_1(x)h_2(x)dx$. We shall freely use certain properties of (h_1,h_2). All of these may be derived in an elementary way from this definition. For vectors, $(h_1,h_2) = \sum_{j=1}^n h_{1j}h_{2j}$.

Equation (5.9) *is* valid (provided an appropriate interpretation of convergence is used)[3] if for some f we can write

(5.10) $$g = Kf.$$

Thus g is in the *range* of K. If there is only a finite number of eigenvalues, the convergence question does not arise. In that case $K(x,y)$ is separable and, as we saw in the previous chapter, all considerations may be reduced to the matrix case. The expansion theorem, and other results that we shall encounter thus lead to facts concerning matrices of which you may not be aware.

Disregarding the constraint of Eq. (5.10) for the moment, let us try to represent $K(x,y)$, considered as a function of x, in a series of the form Eq. (5.9):

$$K(x,y) = \sum_{n=1}^{\infty} a_n(y)\phi_n(x),$$

$$a_n(y) = \int_a^b K(x,y)\phi_n(x)dx = \int_a^b K(y,x)\phi_n(x)dx = \lambda_n \phi_n(y),$$

so that formally

$$K(x,y) = \sum_{n=1}^{\infty} \lambda_n \phi_n(x)\phi_n(y).$$

(Note that the symmetry of K and the orthonormality of the ϕ_n's have been used in obtaining $a_n(y)$.) This expansion is valid if $K(x,y)$ is separable, in which case the sum contains a finite number of terms. It is also valid for a *much* larger class of kernels.

5.4 Eigenvalues and Eigenfunctions of Nonsymmetric Operators—Singular Values and Singular Functions

When we remove the assumption of symmetry, relatively little of the preceding section can be salvaged. First, an integral operator may have *no* eigenvalues at all. For example, if

$$K(x,y) = 0, \qquad y > x,$$

---

[3] The phrase "provided an appropriate interpretation of convergence is used" should be employed almost everywhere infinite series are encountered in this book. I shall refrain from constantly repeating it. See Appendix C if you are interested in various modes of convergence. For the most part, the operators and functions arising in practice allow us to think in terms of ordinary convergence.

Integral Operators with General Kernels

then
$$K\cdot = \int_0^x K(x,y)\cdot dy,$$
called a Volterra operator, has no nonzero eigenvalues. For many such kernels $K(x,y)$ zero is not an eigenvalue. (See Problems 11 and 12.) Fact (1) of Section 5.3 no longer holds.

This is both surprising and worrisome. Matrices *almost always* have some nonzero eigenvalues. If you have thought ahead to the problem of numerical computations, you already probably have realized that matrices are likely to arise when we try to discretize a problem. Such a manipulation may introduce entirely spurious eigenvalues.

A nonsymmetric operator may have eigenvalues and corresponding eigenfunctions. Those eigenvalues may be complex, as may the eigenfunctions. Orthogonality results disappear, as do expansion theorems. If there is an infinite set of eigenvalues, then it is still true that $\lim_{n\to\infty}\lambda_n = 0$

Before further discussion of these matters let us turn to our primary question, the solution of the IFK $g = Kf$. Because g is known, we may apply to it any reasonable operator T that we choose and obtain a known function $h = Tg$. But $Tg = T(Kf) = TKf$. Thus our equation becomes $h = TKf$, where h and the operator TK are known and f is to be determined. (You may detect a few technical problems in this approach, but I choose to ignore them.)

What operator T should we pick to simplify our problem? We choose $T = K^*$, where
$$K^*\cdot = \int_a^b K(x,y)\cdot dx.$$
Note that the variables in $K(x,y)$ have switched roles; the integration is now on the *first* variable. We have *transposed* the variables. Indeed, K^* is quite analogous to the transpose matrix. Observe that if K is symmetric, $K = K^*$. Let us see in full detail what the equation $h = TKf = K^*Kf$ looks like:

$$\begin{aligned} h(z) &= \int_a^b K(x,z)g(x)dx \\ &= \int_a^b K(x,z)dx \int_a^b K(x,y)f(y)dy \\ &= \int_a^b f(y)\left[\int_a^b K(x,z)K(x,y)dx\right]dy. \end{aligned}$$

Thus
$$K^*K\cdot = \int_a^b \left[\int_a^b K(x,z)K(x,y)dx\right]\cdot dy,$$

which is to say that K^*K is the integral operator with kernel

$$K^{(2)}(z,y) = \int_a^b K(x,z)K(x,y)dx.$$

From this explicit representation we readily find that $K^{(2)}(z,y) = K^{(2)}(y,z)$. Therefore K^*K is a *symmetric* operator, and it has all the nice properties listed in Section 5.3.

We shall pursue this approach further when we turn in more detail to IFKs. At this point we use our observations merely as a motivation for the discussion that follows. In that discussion we shall need to recognize that the nonzero eigenvalues of K^*K are not only real, they are positive. We demonstrate this. (The derivation given below can be considerably shortened notationally if you are comfortable with inner products and the like.)

Let $\eta_i \neq 0$ and $v_i(x)$ denote an eigenvalue and corresponding normalized eigenfunction of K^*K:

$$\eta_i v_i(x) = \int_a^b K^{(2)}(x,y)v_i(y)dy.$$

Then

$$\begin{aligned}
\eta_i &= \eta_i \int_a^b v_i^2(x)dx = \int_a^b v_i(x)dx \int_a^b K^{(2)}(x,y)v_i(y)dy \\
&= \int_a^b v_i(x)dx \int_a^b v_i(y)\left[\int_a^b K(t,x)K(t,y)dt\right]dy \\
&= \int_a^b dt \int_a^b K(t,x)v_i(x)dx \int_a^b K(t,y)v_i(y)dy \\
&= \int_a^b dt \left[\int_a^b K(t,x)v_i(x)dx\right]^2 \geq 0.
\end{aligned}$$

The interesting properties of K^*K suggest that we introduce a new operator KK^*. A bit of calculation reveals that in general $KK^* \neq K^*K$. However, KK^* is also symmetric and its eigenvalues are nonnegative. Surprisingly, its eigenvalues are exactly the same as those of K^*K. (See

[11, 46].) The eigenfunctions of the two operators differ in general, a fact that is *not* surprising.

For the positive eigenvalues η_i we write $\sigma_i = \sqrt{\eta_i}$. Thus

$$K^*Kv_i = \sigma_i^2 v_i.$$

Now we define

(5.11) $$u_i = \frac{1}{\sigma_i} Kv_i.$$

Then

(5.12) $$K^*u_i = \frac{1}{\sigma_i} K^*Kv_i = \sigma_i v_i.$$

Notice that Eq. (5.11) and Eq. (5.12) are "dual" relationships. If we apply K to Eq. (5.12), we obtain

$$KK^*u_i = \sigma_i Kv_i = \sigma_i^2 u_i.$$

Thus the u_i defined by Eq. (5.12) are eigenfunctions of the operator KK^*. Moreover, it may be shown that Eq. (5.12) generates *all* such eigenfunctions. (Of course, we may start with the eigenfunctions u_i of KK^* and reverse all steps of the argument to show that the functions v_i defined by Eq. (5.12) are the eigenfunctions of K^*K.)

Summarizing to this point, we have associated with the original operator K two sets of functions v_i and u_i, which are, respectively, eigenfunctions of K^*K and KK^*. They are related by

$$Kv_i = \sigma_i u_i, \qquad K^*u_i = \sigma_i v_i,$$

where the σ_i are the positive square roots of the nonzero eigenvalues of KK^* and K^*K. The functions v_i and u_i are called *singular* functions belonging to K. The σ_i are referred to as *singular* values. (Note that the term "singular" here has nothing to do with continuity.) For a much more complete discussion, see [11, 46].

Finally we note that there may exist functions v such that $K^*Kv = 0$. We include these among the singular functions and agree that the corresponding singular value is zero. It is easy to see that $u = Kv$ satisfies $KK^*u = 0$.

Let us now consider a function g given by $g = Kf$ for some f. Suppose that g can be expanded in a series of the u_i. (Observe that the

u_i's, eigenfunctions of the symmetric operator KK^*, are orthogonal. We assume them to be normalized.)

(5.13) $$g(x) = \sum_{n=1}^{\infty}(g, u_n)u_n(x) = \sum_{n=1}^{\infty}(Kf, u_n)u_n(x).$$

To better understand the coefficients (Kf, u_n) we compute for arbitrary h_1 and h_2

$$\begin{aligned}(Kh_1, h_2) &= \int_a^b (Kh_1)h_2(y)dy \\ &= \int_a^b \left[\int_a^b K(y,t)h_1(t)dt\right] h_2(y)dy \\ &= \int_a^b h_1(t)dt \int_a^b K(y,t)h_2(y)dy \\ &= \int_a^b h_1(t)(K^*h_2)dt \\ &= (h_1, K^*h_2).\end{aligned}$$
(5.14)

You will note that the relationship $(Kh_1, h_2) = (h_1, K^*h_2)$ is quite general, holding for any two functions h_1 and h_2. In the terminology of functional analysis it states that K^* is the *adjoint* operator to K.

Applying Eq. (5.14) to Eq. (5.13) we find, using Eq. (5.12) that

$$\begin{aligned}g(x) &= \sum_{n=1}^{\infty}(Kf, u_n)u_n(x) \\ &= \sum_{n=1}^{\infty}(f, K^*u_n)u_n(x) \\ &= \sum_{n=1}^{\infty}(f, \sigma_n v_n)u_n(x),\end{aligned}$$

so

(5.15) $$g(x) = \sum_{n=1}^{\infty} \sigma_n(f, v_n)u_n(x).$$

This expression is somewhat suspect because we simply assumed a series expansion of the sort Eq. (5.13). Actually, it may be shown that Eq. (5.15) is quite valid for any function g that can be written as $g = Kf$.

It should be noticed that Eq. (5.15) is an apparent generalization of the expansion Eq. (5.9) found in the case of symmetric kernels. There $K = K^*$ and, apparently,

(5.16) $$v_n = u_n = \phi_n, \quad \lambda_n = \sigma_n.$$

Unfortunately Eq. (5.16) is not quite correct because it implies $\lambda_n \geq 0$, which may not be the case. The difficulty lies in the fact that by insisting that σ_n be nonnegative we have selected a particular function pair $\{v_n, u_n\}$. The pair $\{v_n, -u_n\}$ is an equally satisfactory pair of singular functions provided we allow $\sigma_n \leq 0$.

To avoid an involved and not particularly illuminating discussion, we accept Eq. (5.16) and *for the moment* abandon the condition $\sigma_n \geq 0$. (If you are interested in a more detailed treatment, consult [46].) Then, from Eq. (5.15)

$$g(x) = \sum_{n=1}^{\infty} \lambda_n (f, \phi_n) \phi_n,$$

a somewhat more explicit expression than Eq. (5.9).

This result suggests a possible expansion for an arbitrary $K(x,y)$ analogous to that discussed in Section 5.3 for symmetric kernels, namely,

$$K(x,y) = \sum_{n=1}^{\infty} \sigma_n u_n(x) v_n(y).$$

It may be shown that this expansion is indeed valid, under rather mild conditions on K.

Finally, it should be noticed that when K comes from a separable kernel all of these results automatically produce relationships for matrices, and some of these may be new to you. They are all related to the singular value decomposition theory for matrices. The singular vectors and values of square matrices are defined in a manner entirely analogous to the singular functions and values of integral operators, and many of their properties are similar to those we have been discussing. Extensions to nonsquare matrices can be made. For details, see [18].

5.5 Integral Equations of the Second Kind with Symmetric Kernels

We briefly investigate integral equations of the second kind with nonseparable kernels. We confine this investigation to the case of symmetric kernels, those ultimately being the type of primary interest to us in our study of IFKs.

We consider, then

$$(5.17) \quad f(x) = g(x) + \gamma \int_a^b K(x,y)f(y)dy,$$
$$K(x,y) = K(y,x), \quad \gamma \neq 0,$$

which we rewrite in the form

$$f = g + \gamma K f.$$

Notice that this equation implies that $1/\gamma(f-g)$ is representable as Kf provided that Eq. (5.17) has a solution. Therefore, according to the previous section, such a solution f must have the property that

$$f(x) - g(x) = \sum_{n=1}^{\infty} a_n \phi_n(x), \quad a_n = (f-g, \phi_n).$$

From Eqs. (5.17) and (5.14) and the fact that $K = K^*$, we get

$$(f, \phi_n) - (g, \phi_n) = (f-g, \phi_n) = \gamma(Kf, \phi_n) = \gamma(f, K\phi_n) = \gamma\lambda_n(f, \phi_n)$$

so that

$$(f, \phi_n) = \frac{(g, \phi_n)}{1 - \gamma\lambda_n}.$$

Thus

$$a_n = (f-g, \phi_n) = (f, \phi_n) - (g, \phi_n) = (g, \phi_n)\left\{\frac{1}{1-\gamma\lambda_n} - 1\right\}$$
$$= \frac{\gamma\lambda_n}{1-\gamma\lambda_n}(g, \phi_n)$$

and

$$(5.18) \quad f(x) = g(x) + \sum_{n=1}^{\infty} \frac{\gamma\lambda_n}{1-\gamma\lambda_n}(g, \phi_n)\phi_n(x).$$

It should be noted that no assumption concerning the expandability of g in a series of the ϕ's has been necessary.

The expression clearly makes the case $\gamma = \lambda_n^{-1}$, already encountered at the end of Section 4.6, a very special one. No solution f can exist if, for some j, $\gamma = \lambda_j^{-1}$ and $(g, \phi_j) \neq 0$. If $\gamma = \lambda_j^{-1}$ and $(g, \phi_j) = 0$, the corresponding term in Eq. (5.18) becomes meaningless, and we must investigate further. Such a study reveals that Eq. (5.17) does not then have a unique solution. If f is a solution, then so is $f + C_j\phi_j$, where C_j is a completely arbitrary constant.

It must also be recalled that eigenvalues may "repeat," $\lambda_j = \lambda_{j+1} = \cdots = \lambda_{j+m}$, m finite, with corresponding eigenfunctions $\phi_j, \phi_{j+1}, \cdots, \phi_{j+m}$. Suppose we have such repetition and that $\gamma = \lambda_j^{-1}$. Assume also that $(g, \phi_j) = (g, \phi_{j+1}) = \cdots = (g, \phi_{j+m}) = 0$. Then, if f solves Eq. (5.17), so does $f + \sum_k^m C_{k+j}\phi_{k+j}$, where the C's are arbitrary.

Actually we have touched on the Fredholm Alternative. (For more detail, see [11, 33, 46].)

5.6 Integral Equations of the First Kind with General Kernels

We recall Eq. (5.15) which holds if $g = Kf$. In the discussion leading to Eq. (5.15) we supposed f to be known. Now we consider that g is the known function, with f to be determined. It is at once clear that *if* there is a solution f to the IFK $g = Kf$, then g must be of the form Eq. (5.15). We write, assuming the u_n orthonormal,

$$g(x) = \sum_{n=1}^{\infty} g_n u_n(x), \qquad g_n = (g, u_n).$$

The presence of the terms (f, v_n) in Eq. (5.15) suggests that we expand f in a series of the v's:

$$f(y) = \sum_{n=1}^{\infty} f_n v_n(y).$$

Putting this expression in $g = Kf$ then gives (recall Eq. (5.11))

$$\sum_{n=1}^{\infty} g_n u_n = \sum_{n=1}^{\infty} f_n K v_n = \sum_{n=1}^{\infty} \sigma_n f_n u_n,$$

from which we conclude

$$g_n = \sigma_n f_n, \qquad f_n = \frac{(g, u_n)}{\sigma_n}.$$

Hence

(5.19)
$$f(y) = \sum_{n=1}^{\infty} \frac{(g, u_n)}{\sigma_n} v_n(y).$$

The foregoing is, of course, completely formal. In fact, the terms

$$f_n = \frac{(g, u_n)}{\sigma_n}$$

seem particularly distressing, because, if there are infinitely many σ_n, then $\sigma_n \to 0$. It would appear that the f_n get large and the series Eq. (5.19) cannot converge in any reasonable sense. It may be shown that this is false. The constraints that must necessarily be put on g to assure a solution to the IFK imply that the $f_n \to 0$ and that the formal solution Eq. (5.19) is a valid series expansion.

Observe that Eq. (5.19) is *a* solution to the IFK. There is no claim to uniqueness. That is quite another matter, as we saw in Section 5.2.

The worries just experienced about the convergence of the series expansion for f suggest that we return to the preceding section to see if convergence problems occur in the solution to the integral equation of the second kind. There we recall that the corresponding term of interest is

$$\frac{\gamma \lambda_n}{1 - \gamma \lambda_n}(g, \phi_n)$$

(see Eq. (5.18)). If there are infinitely many λ_n then $(1 - \gamma \lambda_n) \to 1$. Thus for large n these terms behave like $\gamma \lambda_n (g, \phi_n)$. The eigenvalue *helps* the convergence rather than inhibiting it. This is a further indication that integral equations of the second kind (at least in the symmetric case) are much easier to deal with than IFKs.

5.7 Summary

In this chapter I have covered, in an informal fashion, some of the properties of integral operators and integral equations that are most pertinent to our studies. We have seen that for the IFK $g = Kf$ to have any solution, g must be adequately restricted. Even then there is no assurance that the equation has a unique solution.

Looking ahead for a moment, we consider what this means with respect to the problem of approximate solution of the IFK when g is known at only a few points and there imperfectly. Suppose, for instance, that

g must be an odd function for a solution to exist. The approximate g is highly unlikely to have this property. How will this reflect on the approximate solution? If we recall the coefficients $(g, u_n)/\sigma_n$ and remember that they are well behaved because of the constraints K automatically imposes upon g, we can see that any "imperfections" in g can create chaos. The terms in the series for f may get large.

In Chapter 6 I shall describe some of the efforts that have been made to resolve the difficulties that are now more obvious. In that chapter we shall have less and less to do with infinite processes, which always have to yield to finite approximations when numerics are introduced.

Problems V

1. Prove that $g_n(x)$ (Eq. (5.4)) has a sup-norm. Also show $\|\psi_n(y)\|_S = \|\cos ny\|_S = 1$.

2. Obtain the result about the unboundedness of the inverse of K by use of the function
$$\psi_n(y) = \begin{cases} 0, & 0 \le y \le y_1; \\ \cos ny, & y_1 < y < y_2; \\ 0, & y_2 \le y \le \pi. \end{cases}$$

3. Repeat Problem 2 without the assumption on $\partial K/\partial y$ by using the Riemann–Lebesgue theorem (Section 5.2).

4. Bessel's inequality (see [32]) states that if $h(x)$ has a two-norm on $0 \le x \le 2\pi$ and h_n are the Fourier cosine coefficients of $h(x)$ then $\sum_{n=0}^{\infty} h_n^2 \le \int_0^{2\pi} h^2(x)dx$. Using this fact, show that if $K(x,y)$ has a two-norm, $0 \le x, y \le 2\pi$, then K does not have a bounded inverse in the space of functions with two-norms. (You may find that a slightly delicate result from integration theory is needed.)

5. Using Bessel's inequality (Problem 4) try to construct sequences b_n and c_{n+1} and a corresponding set of functions
$$\psi_n(y) = b_n \cos ny + c_{n+1} \cos(n+1)y$$
so that ψ_n may be used in the definition of $g_n(x)$ (Eq. (5.4)) instead of just $\cos ny$. Generalize as far as you can, thus showing that the demonstration of unboundedness of K^{-1} is not really so dependent on the choice "$\cos ny$" as it may seem to be.

6. Let $K(x,y)$ be nonseparable but symmetric. Suppose it has two equal eigenvalues λ_1 and λ_2 with corresponding eigenfunctions ϕ_1 and ϕ_2. Show that for any constants a_1 and a_2 $\phi_3 = a_1\phi_1 + a_2\phi_2$ is an eigenfunction with the same eigenvalue. Show that the a's may be chosen so that ϕ_3 is normal.

7. For $K(x,y)$ symmetric, prove that if ϕ_1 and ϕ_2 belong to different eigenvalues then ϕ_1 is orthogonal to ϕ_2. Have you made any assumptions about norms?

8. Prove that for $K(x,y)$ symmetric all eigenvalues are real and that all eigenfunctions may be chosen to be real.

9. Given the symmetric matrix

$$\mathbf{K} = \begin{pmatrix} 2 & 3 \\ 3 & 10 \end{pmatrix},$$

find its eigenvectors and eigenvalues. Solve explicitly

$$\mathbf{g} = \mathbf{K}\mathbf{f}$$

where $\mathbf{g} = \begin{pmatrix} 2 \\ 3 \end{pmatrix}$, using the matrix analogue of Eq. (5.19). Check with standard algebraic methods.

10. Let $K(x,y) = xy + 3x^2y^2$, $0 \le x, y \le 1$. Note that K is separable. Find its eigenfunctions and eigenvalues and demonstrate that

(*) $$K(x,y) = \lambda_1 \phi_1(x)\phi_1(y) + \lambda_2 \phi_2(x)\phi_2(y).$$

Associated with K is a matrix $\mathbf{K}$. Find the analogue of Eq. (*) for $\mathbf{K}$. Try to generalize.

11a. Let $K(x,y)$, $0 \le x, y \le 1$ have a sup-norm. Consider the equation

(**) $$\lambda\phi(x) = \int_0^x K(x,y)\phi(y)dy, \qquad 0 \le x \le 1,$$

where ϕ has a sup-norm on $0 \le y \le 1$. Rewrite this as

$$\lambda^2 \phi(x) = \int_0^x K(x,y)\lambda\phi(y)dy$$

$$= \int_0^x dy K(x,y) \int_0^y K(y,z)\phi(z)dz.$$

Thus

$$|\lambda^2 \phi(x)| \leq \|K\|_S^2 \|\phi\|_S \int_0^x dy \int_0^y dz$$

$$\leq \|K\|_S^2 \|\phi\|_S \sup \left\{\int_0^x dy \int_0^y dz\right\}$$

$$\leq \frac{\|K\|_S^2 \|\phi\|_S}{2}.$$

Iterate this argument to obtain

$$|\lambda^n \phi(x)| \leq \frac{\|K\|_S^n \|\phi\|_S}{n!},$$

and conclude that $\phi(x) \equiv 0$ unless $\lambda = 0$.

When $K(x,y) \equiv 0$ for $y > x$, as is the case in Eq. (**), $K(x,y)$ is called a Volterra kernel. Equations involving such kernels are referred to as Volterra equations. You have demonstrated that under certain conditions the Volterra operator has no eigenvalues except perhaps zero.

11b. If $\lambda = 0$ is an eigenvalue then Eq. (**) becomes

$$(***) \qquad 0 = \int_0^x K(x,y)\phi(y) dy.$$

Let $K(x,y) = (x-y)^n$, where n is zero or a positive integer. Show by repeated differentiation of Eq. (***) with respect to x that $\phi(y) \equiv 0$. Therefore $\lambda = 0$ is not an eigenvalue.

11c. Consider 11b with $n = -\frac{1}{2}$. Use Eq. (2.7) to show that again $\lambda = 0$ is not an eigenvalue.

11d. Both 11b and 11c are special cases of the equation

$$0 = \int_0^x h(x-y)\phi(y) dy.$$

Use the convolution theorem for the Laplace transform to conclude that, in general, $\phi(y) \equiv 0$.

12. Repeat Problem 11a assuming ϕ and K have two-norms.

13. Show that η_i, the eigenvalues of K^*K, are nonnegative, using the inner product notation.

14. Let $K(x,y) = xy^2 + x^3 y$, $0 \leq x, y \leq 1$. Find K^* and $K^{(2)}(x,y)$. Calculate the singular values and singular functions. Solve the equation $g = Kf$, where $g(x) = x + x^3$, using Eq. (5.19).

15. Observe that Problem 14 suggests a solution method for the matrix equation $\mathbf{g} = \mathbf{Kf}$. Try to formulate a general result.

16. From $K^*Kv_i = \sigma_i^2 v_i$ obtain $K(K^*Kv_i) = \sigma_i^2 Kv_i$. Set $Kv_i = \tilde{\phi}_i$ so that $KK^*\tilde{\phi}_i = \sigma_i^2 \tilde{\phi}_i$. Hence reason that K^*K and KK^* have the same eigenvalues. Do you see any difficulties?

17. Return to Problem 14. Verify that $K(x,y) = \sum_{n=1}^{2} \sigma_n u_n(x) v_n(y)$.

18. Given $K(x,y) = xy + 1$, $0 \leq x, y \leq 1$, solve $f = g + \gamma Kf$, where $g(x) = x^2$, by use of Eq. (5.18).

19. Consider the following approach to solving Eq. (5.17). Set $f_0(x) = g(x)$, $f_1(x) = g(x) + \gamma \int_a^b K(x,y) f_0(y) dy, \cdots,$ $f_n(x) = g(x) + \gamma \int_a^b K(x,y) f_{n-1}(y) dy$, $n = 2, 3, \cdots$. Assume K and g have sup-norms. Show that if γ satisfies a certain restriction, then $\lim_{n \to \infty} f_n(x) = f(x)$ exists and solves Eq. (5.17). Have you used any symmetry properties of $K(x,y)$?

Note that $f(x)$ is given by a series called a Neumann series. Write this series explicitly.

20. In Problem 19 suppose $K(x,y)$ is Volterra (see Problem 11). Show that in this case $\lim_{n \to \infty} f_n(x) = f(x)$ for *all* γ.

21. The difficulties which exist when $\gamma = \lambda_j^{-1}$ in Eq. (5.17) are clearly demonstrated by Eq. (5.18). Show that if $(g, \phi_1) = 0$, Eq. (5.17) has a nonunique solution.

Generalize to the case in which λ_j "repeats."

CHAPTER 6

Some Methods of Resolving Integral Equations of the First Kind

6.1 Introduction

Our discussion thus far has emphasized the difficulties inherent in IFKs. My purpose has not been to discourage you, but to make you aware that the problems you have in trying to solve an IFK in a practical context are usually not of *your* making but are direct reflections of the very unpleasant properties of the equations themselves. Now that these matters are at least partially clarified, I tackle the question of how to resolve IFKs.

As noted earlier the term "resolve" seems more appropriate than "solve." Under our basic assumption that the data are known only at a finite number of points, and there only inexactly, a solution in the usual sense is generally impossible. It has become clear that the kernel K forces g to have certain properties. The data may not accurately reflect those properties. This suggests the possibility of modifying g, but distortion of (presumably) measured data involves many dangers.

A strong argument may often be made for restricting the class of f's that can be possible solutions to a problem. For example, the physical origins may demand that f be nonnegative, or that it be monotonic. Such restrictions, based on solid knowledge, can validly be imposed. However, there is often a strong tendency to ascribe behavior to f that we only *hope* it possesses—it does not oscillate too badly, it has a single maximum, it has a certain behavior near a or b, etc. Very seldom does the structure of the IFK impose these conditions. They are usually quite subjective, and there may be a fine line between genuine physical intuition (or divine inspiration) and poorly based faith or hope.

It is clear from the unboundedness of K^{-1} and the fact that small changes in g may produce "wild" behavior in f, that means must often be found to "tame" this solution function. Researchers should be quite aware that they have a great deal of control over and responsibility for this taming. Should they guess that the solution is monotonic, although it is actually oscillatory, they may obtain a "solution" that has completely incorrect properties.

The necessity of such subjective constraints leads to the (nonstandard) terminology *resolve* in place of solve. There is nothing wrong with doing this sort of thing. It actually is employed very commonly in other numerical work, often in a very subtle way. Here it is important to realize what is being done and why. The employment is not subtle, it is overt. But the IFK is not subtle either.

You undoubtedly realize that we must now resort to numerical methods and you perhaps expect this chapter to include a plethora of codes and programs. This is not the case. I shall emphasize concepts and ideas. When a specific numerical device is presented, a very simple computational scheme will generally be used to generate examples. No codes will be given. Some of the reasons for this will become apparent as we go along. The matter will be discussed further in Chapter 7.

This chapter is quite long and is broken into sections and subsections. Each of these sections contains a method of resolution. I try to point out those methods that seem to show greatest promise and those that are less satisfactory. Such a wide collection of algorithms exists in the literature that it is impossible to touch upon them all. I discuss only the more common and more successful ones.

6.2 Classical Quadrature Approach

6.2.1 Basic Scheme

It has been mentioned earlier that the very structure of an IFK suggests a matrix approach. Let us suppose the data are known at points x_i, $i = 1, 2, \cdots, N$. Thus

$$g(x_i) = \int_a^b K(x_i, y) f(y) dy.$$

Resolving Integral Equations of the First Kind

Now we replace the integration by some quadrature scheme. For convenience we suppose an N-point quadrature formula. We also write equality when actually an approximation has already been made. Thus

$$g(x_i) = \sum_{j=1}^{N} K(x_i, y_j) w_j f(y_j),$$

where the w's are the quadrature weights.

It is immediately apparent that we now have a matrix problem:

(6.1) $$\mathbf{g} = \mathbf{K}\mathbf{f},$$

where

$$\begin{aligned}(\mathbf{K})_{ij} &= K(x_i, y_j) w_j = k_{ij}, \\ \mathbf{g} &= (g(x_1), g(x_2), \cdots, g(x_N))^T, \\ \mathbf{f} &= (f(y_1), f(y_2), \cdots, f(y_N))^T.\end{aligned}$$

What constraint has been placed upon the solution function? It has been replaced by a function that assumes values at only discrete values y_j. Although we might consider it to be undefined elsewhere, it is probably a little more instructive to think of it as a step function or histogram (see Fig. 6.1).

If we now assume that $\mathbf{K}$ has an inverse we can at once obtain $\mathbf{f}$:

$$\mathbf{f} = \mathbf{K}^{-1}\mathbf{g}.$$

To see what can go wrong with this approach, we consider in the next section a particular problem.

6.2.2 A Particular Application of the Quadrature Approach

Let us consider the IFK

$$g(x) = \int_0^1 e^{-\alpha|x-y|} f(y) dy.$$

If

$$g(x) = \frac{2x}{\alpha} - \frac{e^{\alpha(1-x)}}{\alpha} + \frac{1}{\alpha^2}[e^{-\alpha x} - e^{-\alpha(1-x)}],$$

the (unique) solution is simply

$$f(y) = y,$$

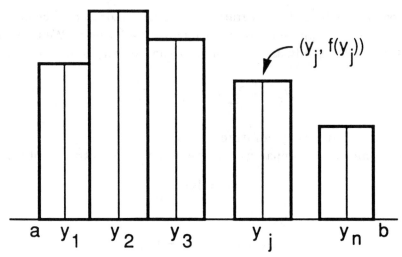

Figure 6.1. *Histogram for f*.

as may be seen from Section 2.7.

The scheme outlined in the previous section was implemented for this problem using simple trapezoidal quadrature and equally spaced points. The data were evaluated at the quadrature points to machine accuracy (about 14 decimal digits) and the matrix **K** was calculated (for various values of N) for the case $\alpha = 5.0$. The system Eq. (6.1) was then solved using full machine accuracy. For $N = 6, 21$, and 101, the results agreed with the true solution to the accuracy expected from the trapezoidal rule.

"Inaccuracy" was then introduced into the data by writing

$$g_e(x_i) = g(x_i) + \epsilon(i), \qquad \epsilon(i) = \bar{\epsilon}\mathrm{Rand}(i),$$

where Rand(i) is a uniformly distributed random variable on $[-1, 1]$. All calculations were repeated using g_e instead of g. The cases $\bar{\epsilon} = 10^{-4}$ and $\bar{\epsilon} = 10^{-2}$ were investigated.

Obviously, a quantitative measure of the error in the solution is desirable. Many such criteria may be devised. Here and in all subsequent work, I define

(6.2) $$\rho = \frac{1}{\sqrt{N}}\left\{\sum_{j=1}^{N}(f_c(y_j) - f(y_j)^2\right\}^{1/2},$$

where f_c denotes the computed solution and f is the true solution. (Clearly, $f(y_j) = y_j$ in the example under consideration.) Thus we will be using the root mean square error as our error criterion throughout our work.

The values of ρ for the experiments described are given in Table 6.1. (The symbol (n) denotes 10^n.)

The results for $\bar{\epsilon} = 0$ are not surprising. We anticipate improvement as the number of quadrature points increases. For $\bar{\epsilon} = 10^{-4}$ there is improvement from $N = 6$ to 21 and then serious degradation when N is increased to 101. For the larger error $\bar{\epsilon} = 10^{-2}$, increasing N evidently simply makes matters worse. Clearly, for noisy data we should use only a few quadrature points. But then we suffer a mean square error of about 0.06 which is 6% of the maximum value of $f = y$, $0 \leq y \leq 1$.

Immediately the possibility of a better quadrature scheme comes to mind. Indeed, the relative insensitivity of ρ to the error size in the case $N = 6$ suggests that most difficulty there probably does come from the crudity of the trapezoidal rule. But experiments with other numerical schemes (see Problem 1) will reveal that the overall behavior found in Table 6.1 will not change.

What has gone wrong? We have replaced the unbounded inverse of an integral operator by the bounded inverse of a matrix operator. We have simultaneously constrained f. As N increases, the inverse matrix operator tries to mimic an unbounded operator. An inspection of the matrix $\mathbf{K}^{-1}$ shows that it does this by becoming badly behaved as N increases, with larger and larger positive and negative entries occurring more or less in alternate locations. When a small error is introduced in g, that error is greatly magnified by multiplication by these entries. Although "lucky" cancellation may occur, it usually does not.

To get a better feeling for what is happening, we digress to another problem and study the case in which $N = 2$. There a transparent anal-

Table 6.1.
ρ Values for Computed Solution

N	$\bar{\epsilon} = 0$	$\bar{\epsilon} = 10^{-4}$	$\bar{\epsilon} = 10^{-2}$
6	0.69 (−1)	0.69 (−1)	0.59 (−1)
21	0.66 (−2)	0.97 (−2)	0.55 (0)
101	0.50 (−3)	0.13 (0)	0.13 (2)

ysis is possible and a better understanding of what is occurring can be obtained. We shall return to the example of this section shortly.

6.2.3 A Trivial Example and Some Important Observations

To better understand the phenomena described in the preceding section we consider the problem

$$(6.3) \qquad 2 = g(x) = \int_{-1}^{1} (1 + \eta x y) f(y) dy, \qquad \eta = \text{constant}.$$

Because the kernel is separable, a solution is very easy to obtain (observe that it is not unique):

$$f(y) = 1.$$

We propose to use an approximate quadrature method with $N = 2$, $x_1 = -1/2$, $x_2 = 1/2$, $y_1 = -1/2$, $y_2 = 1/2$, $w_1 = w_2 = 1$. Thus

$$g_1 = g(-1/2) = f(-1/2)(1 + \eta/4) + f(1/2)(1 - \eta/4),$$

$$g_2 = g(1/2) = f(-1/2)(1 - \eta/4) + f(1/2)(1 + \eta/4),$$

$$\mathbf{K} = \begin{pmatrix} 1 + \eta/4 & 1 - \eta/4 \\ 1 - \eta/4 & 1 + \eta/4 \end{pmatrix}.$$

We readily calculate

$$\mathbf{K}^{-1} = \frac{1}{\eta} \begin{pmatrix} 1 + \eta/4 & -(1 - \eta/4) \\ -(1 - \eta/4) & 1 + \eta/4 \end{pmatrix}$$

so that with

$$f(-1/2) = f_1, \qquad f(1/2) = f_2,$$

$$\begin{pmatrix} f_1 \\ f_2 \end{pmatrix} = \frac{1}{\eta} \begin{pmatrix} 1 + \eta/4 & -(1 - \eta/4) \\ -(1 - \eta/4) & 1 + \eta/4 \end{pmatrix} \begin{pmatrix} g_1 \\ g_2 \end{pmatrix}.$$

For the special case $g_1 = g_2 = 2$ (see Eq. (6.3)), we get

$$f_1 = f_2 = 1.$$

It is interesting to notice that this solution agrees with the analytic solution given. I do not pursue this matter.

Now let us suppose that g is in error and replace g_i with $g_i + \epsilon_i$, where ϵ_i is small. Then

$$\begin{pmatrix} f_{1,\epsilon} \\ f_{2,\epsilon} \end{pmatrix} = \frac{1}{\eta} \begin{pmatrix} 1+\eta/4 & -(1-\eta/4) \\ -(1-\eta/4) & 1+\eta/4 \end{pmatrix} \begin{pmatrix} g_1 + \epsilon_1 \\ g_2 + \epsilon_2 \end{pmatrix}$$

$$= \begin{pmatrix} f_1 \\ f_2 \end{pmatrix} + \frac{1}{\eta} \begin{pmatrix} 1+\eta/4 & -(1-\eta/4) \\ -(1-\eta/4) & 1+\eta/4 \end{pmatrix} \begin{pmatrix} \epsilon_1 \\ \epsilon_2 \end{pmatrix}.$$

Thus the first component of f is in error by

$$\frac{\epsilon_1 - \epsilon_2}{\eta} + \frac{\epsilon_1 + \epsilon_2}{4},$$

and the error in the second component is

$$\frac{\epsilon_2 - \epsilon_1}{\eta} + \frac{\epsilon_1 + \epsilon_2}{4}.$$

Note that these errors do not depend upon the assumption that $g_i = 2$. They depend only upon the structure of **K** and the error in g.

Because ϵ_i is small, the errors in **f** will be small *unless* η is also small. If that is the case, f_ϵ will bear little resemblance to **f**, unless there is some very fortunate cancellation.

Despite the simplicity of this example, quite a bit of insight into what goes on in more realistic cases can be obtained with a little careful analysis. First, we examine **K** and note that for η small the columns (and rows) are very nearly equal. Thus **K** is "almost singular." The elements of $\mathbf{K}^{-1}$ are "alternately" large positive and negative numbers.

Next, we examine the original kernel $K(x,y) = 1 + \eta xy$. For η small the surface represented by this function is very flat. As η increases this flatness disappears. Now, in much of mathematics, functions that change slowly are highly desirable. That is not the case so far as kernels of IFKs are concerned. In fact, if we disregard one of our ground rules for a moment and search for a very nonflat kernel, we recognize that the δ-function, is what we need. Suppose we have $K(x,y) = \delta(x-y)$. Then

$$g(x) = \int_0^1 \delta(x-y) f(y) dy$$

has the obvious solution $f(x) = g(x)$. Kernels that are δ-like are actually highly desirable.

6.2.4 Section 6.2.2 Revisited

The remarks just made concerning δ-like behavior suggest that we examine the kernel $K = e^{-\alpha|x-y|}$ for various values of α. For α small this surface is very flat. For $\alpha = 5$, the case considered, the surface has a pronounced ridge along the line $y = x$. As α increases this ridge becomes sharper. Thus $e^{-\alpha|x-y|}$ becomes more and more δ-like.

To test our somewhat intuitive arguments of the previous Section we repeat the calculations of Section 6.2.2 using $\alpha = 0.5$ and $\alpha = 10.0$. The values of ρ as defined by Eq. (6.2) are given in Table 6.2, where we also include the results for $\alpha = 5.0$ for ease of reference. Again, some of the anomalies in Table 6.2 may be attributed to the use of the trapezoidal rule. (For fixed x and large α, $e^{-\alpha|x-y|}$ is not well approximated by a few straight line segments.) However, results when errors are present are generally better for larger values of α, exactly as anticipated.

6.2.5 A Return to the Kernel $K = 1 + \eta xy$

In Section 6.2.3 we fixed N at the value 2. This was reasonable because the kernel is trivially separable and the integral operator is equivalent to a matrix operator of order 2. Let us ignore that fact for a moment and write the Nth order matrix **K** arising from obvious extension of the quadrature employed in 6.2.3, assuming the quadrature points y_i to be equally spaced.

Table 6.2.
ρ Values for the Computed Solution

N	α	$\bar{\epsilon} = 0$	$\bar{\epsilon} = 10^{-4}$	$\bar{\epsilon} = 10^{-2}$
6	0.5	0.39 (−1)	0.38 (−1)	0.33 (0)
	5.0	0.69 (−1)	0.69 (−1)	0.58 (−1)
	10.0	0.16 (0)	0.16 (0)	0.14 (0)
21	0.5	0.52 (−2)	0.56 (−1)	0.54 (1)
	5.0	0.66 (−2)	0.97 (−2)	0.55 (0)
	10.0	0.14 (−1)	0.15 (−1)	0.29 (0)
101	0.5	0.47 (−3)	0.13 (1)	0.13 (3)
	5.0	0.50 (−3)	0.13 (1)	0.13 (2)
	10.0	0.71 (−3)	0.65 (−1)	0.66 (1)

$$\mathbf{K} = \frac{1}{N} \begin{pmatrix} \cdot & \cdot & \cdot & & \cdot & & \cdot \\ \cdot & \cdot & & \cdot & & \cdot & \cdot \\ \cdot & \cdot & \cdot & & & & \cdot \\ \cdot & \cdot & 1+\eta x_i y_j & 1+\eta x_i y_{j+1} & \cdot \\ \cdot & \cdot & 1+\eta x_{i+1} y_j & 1+\eta x_{i+1} y_{j+1} & \cdot \\ \cdot & \cdot & 1+\eta x_{i+2} y_j & 1+\eta x_{i+2} y_{j+1} & \cdot \\ \cdot & \cdot & & \cdot & & \cdot & \cdot \\ \cdot & \cdot & & & \cdot & & \cdot \\ \cdot & \cdot & & & \cdot & & \cdot \end{pmatrix}$$

For N large, y_{j+1} is very close to y_i. Thus the two columns shown are nearly equal, $\mathbf{K}$ is almost singular, and $\mathbf{K}^{-1}$ can be expected to exhibit poor behavior.

A moment's thought reveals that the particular K under discussion has very little to do with this observation. If $K(x,y)$ is any kernel with reasonable continuity properties, $\mathbf{K}$ will be nearly singular for large N and all the unpleasant behavior we have been discussing will occur.

From our analysis of the simple example of this section we are led to two important bits of knowledge:

1. The flatter is the surface represented by $K(x,y)$ the harder will the IFK be to resolve.

2. For any reasonable K the quadrature method must produce poor results when N is allowed to become large.

6.2.6 The Condition Number of a Matrix

Is there some way of predicting "good" or "bad" behavior for $\mathbf{K}$? The so-called condition number of $\mathbf{K}$ is an answer. We define

(6.4) $$C = \text{condition number of } \mathbf{K} = \frac{\sigma_1}{\sigma_N},$$

where σ_1 and σ_N are the largest and smallest singular values of $\mathbf{K}$. Let $\delta \mathbf{g}$ and $\delta \mathbf{f}$ be the *relative* errors in the data and in the solution. Then it may be shown that (see [18, 47])

(6.5) $$\| \delta \mathbf{f} \|_2 \leq C \| \delta \mathbf{g} \|_2 .$$

Although this provides only an upper bound on $\delta \mathbf{f}$, experience indicates that very often the 2-norm of the error $\delta \mathbf{f}$ is actually very close to the upper bound. Thus, roughly speaking, the solution error can be estimated by multiplying the data error by the condition number.

Many singular value decomposition programs automatically compute C, or an approximation of it (see [18]). (Caution: Some authors and some software use $\frac{1}{C}$ as the condition number.)

6.2.7 A Digression Concerning Approximate Kernels

It was emphasized in Chapter 1 that $K(x, y)$ will always be assumed to be completely known. However, in the preceding sections K has been approximated. The nature of that approximation can be understood most easily by examining the quadrature used in Section 6.2.3. There the given K has been replaced by a function that is constant in each of the four quadrants of the square $-1 \leq x, y \leq 1$. The function obtained in the more general quadrature of Section 6.2.5 may be similarly interpreted. This quadrature replaces the original K by a "post-pile" function.

Now a post-pile kernel can be written as a separable kernel,[1]

$$K(x,y) = \sum_{i=1}^{N} \sum_{j=1}^{N} k_{ij} \tilde{\alpha}_i(x) \tilde{\beta}_j(y),$$

by appropriate choice of $\tilde{\alpha}_i$ and $\tilde{\beta}_j$. Thus the original kernel has been replaced by a separable kernel that approximates it.

A basic question that should be asked is: If $K(x,y)$ is approximated, will the corresponding singular values, singular functions and solutions of integral equations of both first and second kinds be similarly approximated? The answer is a cautious yes. In fact, one of the standard ways of obtaining information about integral operators is by replacing them with operators corresponding to approximating separable kernels. Obviously, some sort of limiting process is required in applying such a technique. This limiting process is often delicate, a further reflection of the unpleasant behavior of the inverse of an integral operator.

---

[1] Note that this function may be put in the more standard form by writing $\sum_{j=1}^{N} k_{ij} \tilde{\beta}_j(y) = \beta_i(y), \tilde{\alpha}_i(x) = \alpha_i(x)$.

To pursue this matter analytically would take us too far afield. A numerical example may be helpful. We consider the sequence of kernels

$$K_n(x,y) = \sqrt{xy} \sum_{j=0}^{n} \frac{(-1)^j (xy)^j}{(2j+1)!}, \qquad 0 \le x, y \le 1,$$

and observe that

$$\lim_{n \to \infty} K_n(x,y) = \sin \sqrt{xy} = K(x,y).$$

Because the singular values and singular functions (in this case actually the eigenvalues and eigenfunctions) are completely determined by the kernel and, in turn, completely determine the kernel (see Chapter 5) it suffices to study those values and functions.

The kernel K_n is separable. Thus it has only n nonzero eigenvalues. Those have been calculated, together with the corresponding eigenfunctions Then K was replaced by a matrix in the manner of Section 6.2.5. The results are given in Table 6.3 and Figs. 6.2–6.4. Observe that $u_i \equiv v_i$ because of the symmetry. It will be noted that the agreements are excellent. (You are urged to do further experiments of this kind. See Problem 4.)

In the material that follows, the sense in which a given kernel is approximated will not usually be mentioned. Suffice it to say that the overall technique of kernel approximation is acceptable. It is usually the problem of working with these approximations that causes the distress.

6.2.8 The Method of Least Squares

The requirement that the number of data points and the number of quadrature points be equal should be examined briefly. If this is not so, then the method of least squares suggests itself. Taking N_1 as the

Table 6.3. Singular Values σ_j

j	K_0	K_2	K_4	K
1	0.5000(0)	0.4643	0.4643	0.4643
2	0	0.4631(−2)	0.4639(−2)	0.4639(−2)
3	0	0.1495(−4)	0.1391(−4)	0.1391(−4)
4	0	0	0.2023(−7)	0.2026(−7)
5	0	0	0.1860 (−10)	0.1736(−10)

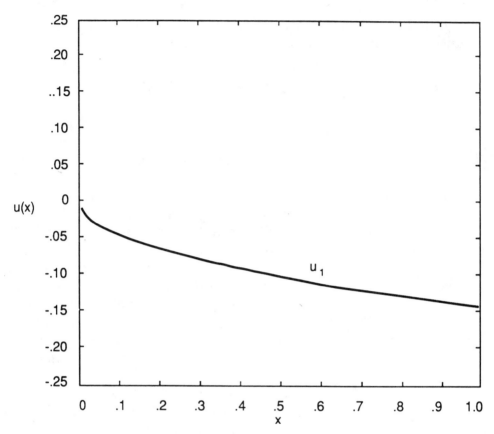

Figure 6.2. *Singular function $u_1(x)$ for K_0.*

number of data points and N_2 ($N_2 < N_1$) as the order of the quadrature leads readily to the condition

$$F(f_1 f_2, \cdots, f_{N_2}) = \sum_{i=1}^{N_1} (g_i - \sum_{j=1}^{N_2} k_{ij} f_j)^2 = \text{minimum},$$

$$\frac{\partial F}{\partial f_p} = 2 \sum_{i=1}^{N_1} \left(g_i - \sum_{j=1}^{N_2} k_{ij} f_j \right) k_{ip} = 0,$$

$$p = 1, 2, \cdots, N_2.$$

These equations may be rewritten in matrix form

(6.6) $$\mathbf{K}^* \mathbf{g} = \mathbf{K}^* \mathbf{K} \mathbf{f},$$

Resolving Integral Equations of the First Kind

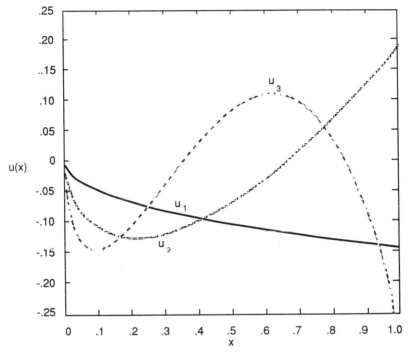

Figure 6.3. *Singular functions $u(x)$ for K_2.*

where K^* is the transpose of K, the analogue of the adjoint of the integral operator.

The matrix K^*K is, of course, square and symmetric. Although it enjoys some agreeable properties (see Section 5.3 for the corresponding properties of the integral operator), Eq. (6.6) involves just about the same difficulties as were encountered in previous sections. Although least squares often may be appropriate in a particular problem, all of the unpleasant behavior discussed already may be anticipated.

6.2.9 Some Miscellaneous Comments Concerning the Quadrature Method

The quadrature method often gives some information concerning solutions to IFK. You should just not expect too much of it. Whether or not it will enjoy some level of success can be determined if an estimate of the error in the data is available. In that case calculation of the condition number of K is also called for.

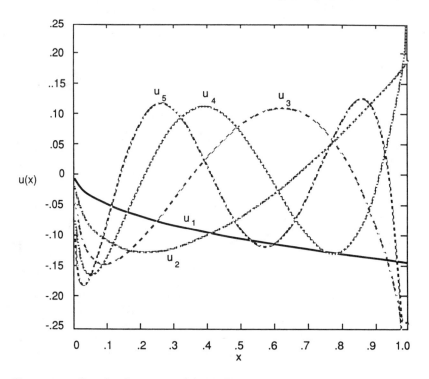

Figure 6.4. *Singular functions $u(x)$ for K_4 and first five singular functions for K.*

It is also possible to incorporate in the quadrature method some of our subjective ideas about the nature of a solution to an IFK. If, for example, the solution is thought to be slowly changing for certain values of the variable and rapidly changing in other intervals, a quadrature scheme may be picked that involves more detail where the rapid changes are expected. This approach may be semiautomated. We may start with little or no idea of the nature of the solution and use a relatively simple quadrature. From the solution thus obtained we may surmise where the function has the most interesting behavior and revise the quadrature method accordingly. Obviously, the scheme may be repeated.

The above idea clearly has many possible variations. It is vital to remember that any attempt to wring more information out of a problem than it can provide (the "classical" attempt involves simply increasing the quadrature order) is doomed.

6.2.10 A Summary of the Classical Quadrature Approach

We have briefly examined what is probably the most obvious device for the approximate solution of IFKs. It has been found that the approach is capable of providing an approximate solution if we do not want too much detail. Although a large number of variants of the quadrature method may be devised to make it more useful, the algorithm has a "natural limit" beyond which it provides unsatisfactory information. This "natural limit" is very problem dependent.

6.3 General Series Expansions

6.3.1 Basic Scheme

The concept of expanding a function into a series is so common in mathematics and physics that it may very well be the next possibility to come to mind. We begin by choosing a set of (more or less) arbitrary functions $\psi_n(x)$ defined on $a \le x \le b$. It is convenient to suppose them already orthonormalized. Write

$$g(x) = \sum_{n=1}^{\infty} g_n \psi_n(x), \qquad f(y) = \sum_{n=1}^{\infty} f_n \psi_n(y),$$

so that the IFK $g = Kf$ becomes formally

$$\sum_{n=1}^{\infty} g_n \psi_n(x) = \sum_{n=1}^{\infty} f_n \int_a^b K(x,y) \psi_n(y) dy.$$

We truncate, but continue to write equality:

$$\sum_{n=1}^{N} g_n \psi_n(x) = \sum_{n=1}^{N} f_n \int_a^b K(x,y) \psi_n(y) dy.$$

In the usual way

$$g_p = \sum_{n=1}^{N} f_n \int_a^b \int_a^b K(x,y) \psi_n(y) \psi_p(x) dx dy = \sum_{n=1}^{N} k_{pn} f_n,$$

and once again the problem has been reduced to one involving matrices:

$$\mathbf{g} = \mathbf{Kf}.$$

There is quite a close connection between this method and that of the previous section. Suppose we pick a very special (finite) set of ψ_n:

$$\psi_n(x) = \frac{1}{\sqrt{x_n - x_{n-1}}}; \qquad x_{n-1} \leq x < x_n,$$
$$\psi_n(x) = 0 \quad \text{elsewhere}, \quad n = 1, 2, \cdots, N, \quad x_0 = a, \quad x_N = b.$$

It is clear that this set is orthonormalized. The coefficients g_p now are given by

$$\begin{aligned} g_p &= \sum_{n=1}^{N} f_n \int_{x_{p-1}}^{x_p} \int_{y_{n-1}}^{y_n} K(x,y) \psi_n(y) \psi_p(x) dx\, dy \\ &= \sum_{n=1}^{N} f_n (\sqrt{x_p - x_{p-1}} \sqrt{y_n - y_{n-1}})^{-1} \int_{x_{p-1}}^{x_p} \int_{y_{n-1}}^{y_n} K(x,y) dx\, dy \\ &= \sum_{n=1}^{N} f_n \tilde{k}_{pn}, \qquad y_n = x_n, \quad n = 1, 2, \cdots, N. \end{aligned}$$

The function $g(x)$ has been replaced by a linear combination of step functions whose value on $x_{n-1} \leq x \leq x_n$ is just $g_n/\sqrt{x_{n-1} - x_n}$; f has also been replaced by a similar expression. The kernel K has been approximated by a post-pile function

$$K(x,y) = \tilde{k}_{pn} \ (x_p - x_{p-1})^{-1/2} (y_n - y_{n-1})^{-1/2},$$
$$x_{p-1} \leq x < x_p, \qquad y_{n-1} \leq y < y_n.$$

Thus for such a set of ψ_n the series method is basically equivalent to a quadrature method.

Of course, the series method allows a considerably greater flexibility than the quadrature approach because more general ψ_n may be chosen. In fact, there is no reason that g and f should be expanded in the same set of functions. We know that g has to be restricted for the IFK to have a solution. It should be expanded in a series that reflects such constraints. Furthermore, any information that we have about f should be somehow contained in its series expansion. Finally, there is no reason to truncate the two series at the same point; the two truncated series can well have different numbers of terms. Incorporating all of these ideas we get, using orthonormal sets ψ_n and θ_n,

$$\sum_{n=1}^{N_1} g_n \psi_n(x) = \sum_{n=1}^{N_2} f_n \int_a^b K(x,y) \theta_n(y) dy,$$

Resolving Integral Equations of the First Kind

so that
$$g_p = \sum_{n=1}^{N_2} f_n \int_a^b \int_a^b K(x,y)\theta_n(y)\psi_p(x)dx\,dy, \qquad p = 1, 2, \cdots, N_1,$$

or, in matrix form,
$$\mathbf{g} = \mathbf{Kf},$$

where **K** is no longer square. This algebraic problem can be resolved by the method of least squares. All of the algebra may involve the same considerations of nearly singular matrices, etc., encountered in Section 6.2. In the present case, however, it is a bit more difficult to predict, through continuity considerations of $K(x,y)$, how $\mathbf{K}^*\mathbf{K}$ will behave. This is because the integrations necessary to obtain the elements of **K** partially conceal the "local" behavior of $K(x,y)$. It is possible to demonstrate, again using condition number arguments, that $\mathbf{K}^*\mathbf{K}$ becomes more and more ill behaved as additional terms in the series expansions are used.

6.3.2 Series Expansions and Projection and Collocation Methods

A slightly different view of what we have just done may be helpful. Assume θ_n and ψ_n are as in the preceding section. Write

$$g(x) = \sum_{n=1}^{N_2} f_n \int_a^b K(x,y)\theta_n(y)dy.$$

Now consider the space spanned by the ψ_n. Project g into this space. That is form $(g, \psi_p) = \int_a^b g(x)\psi_p(x)dx$, $p = 1, 2, \cdots, N_1$:

$$(g, \psi_p) = g_p = \sum_{n=1}^{N_2} f_n \int_a^b \int_a^b K(x,y)\theta_n(y)\psi_p(x)dx\,dy.$$

This is the equation already obtained, using the usual orthonormality arguments.

Next, relax the condition that the ψ_n and θ_n are orthonormal, and require merely that the ψ_n be linearly independent and that the θ_n also be linearly independent. Expand f in a (finite) series of the θ_n, and proceed as above. Again we get

$$(g, \psi_p) = \sum_{n=1}^{N_2} f_n \int_a^b \int_a^b K(x,y)\theta_n(y)\psi_p(x)dx\,dy.$$

No longer can the (g, ψ_p) be thought of as Fourier coefficients. They are simply numbers that can be calculated. As usual, we can solve for the f_n and thus find the (approximate) solution $f(x)$.

This approach is referred to as a *projection* method. Often the term Galerkin method is used in recognition of the Russian engineer who popularized this device. You will note that the difference between the orthogonal series approach and that of Galerkin is to a considerable extent conceptual. The latter has the advantage of not requiring orthonormality, and allows ready employment of such functions as B-splines, now so common in numerical analysis.

We also observe that the projection approach may make good choices of the ψ_n and θ_n somewhat clearer. If, for example, g must be an odd function, it is unwise to use any even ψ_n; g will be orthogonal to such a function. The space spanned by the ψ_n should be as close as possible to the space in which g must lie. Clearly, similar statements pertain to the θ_n, although less is known about the function f, of course.

A very important algorithm can be obtained if we temporarily relax our requirement that the Dirac δ-function not be used. Define

$$\psi_n(x) = \delta(x - x_n), \qquad a \leq x_1 < x_2 \cdots < x_{N_1} \leq b.$$

Then projection gives, formally,

$$\begin{aligned}(g, \psi_p) &= \int_a^b g(x)\delta(x - x_p)dx = g(x_p) \\ &= \sum_{n=1}^{N_2} f_n \int_a^b \int_a^b K(x,y)\theta_n(y)\delta(x - x_p)dx\, dy,\end{aligned}$$

so

$$g(x_p) = \sum_{n=1}^{N_2} f_n \int_a^b K(x_p, y)\theta_n(y)dy.$$

Clearly, the matrix **K** can be calculated with a single integration for each element rather than a double integration, often a great savings in computing time. This approach is called *collocation*.

For a discussion of important properties of Galerkin and collocation methods, see [3, 8, 15, 16].

6.4 Expansion in Series of Eigenfunctions and Singular Functions

6.4.1 Basic Scheme

In the preceding section the function used to expand f and g (equivalently, the spaces into which f and g have been projected) have been chosen primarily on the basis of the behavior of the data and expected behavior of the solution. Properties of the integral operator K have played a secondary role. It is reasonable to ask whether K should not be more fully involved.

We have observed earlier that the operator K determines completely its singular functions and singular values and is, in turn, completely determined by them. (Of course, if K is symmetric, we may talk in terms of eigenvalues and eigenfunctions.) It seems reasonable to use the singular functions for the series expansions. In fact, this has already been discussed in Section 5.6 and an expansion for f has been found:

$$(6.7) \qquad f(y) = \sum_{n=1}^{\infty} \frac{(g, u_n)}{\sigma_n} v_n(y).$$

We have noted that the coefficients $(g, u_n)/\sigma_n$ are a potential source of difficulty. Although they must become small if g is exact, any inaccuracies in g can prevent this. Indeed, suppose that g is replaced by $g + \epsilon$, where ϵ is the data error. Then

$$\frac{(g + \epsilon, u_n)}{\sigma_n} = \frac{(g, u_n)}{\sigma_n} + \frac{(\epsilon, u_n)}{\sigma_n},$$

and in general $(\epsilon, u_n)/\sigma_n$ becomes large as n increases.

This observation suggests that the infinite series in Eq. (6.7) must be truncated. Instead of pursuing this train of thought, we should first recognize that in a practical situation we ordinarily are not able to obtain the full series Eq. (6.7) anyway, nor can we find the v_n, u_n, and σ_n exactly. They must be approximated. We may do this by first replacing the integral operator by a matrix operator. This can be done, for example, by approximating $K(x, y)$ by a separable kernel. (Recall that quadrature is included in this device.)

Once a matrix operator has been obtained, the expansion Eq. (6.7) must be replaced by (see Problems 7 and 8)

$$\mathbf{f} = \sum_{n=1}^{N} \frac{(\mathbf{g}, \mathbf{u}_n)}{\tilde{\sigma}_n} \mathbf{v}_n,$$

where $\mathbf{v}_n$, $\mathbf{u}_n$, and $\tilde{\sigma}_n$ are the singular vectors and singular values of $\mathbf{K}$. (See [27, 47].) Although the series is now finite, the difficulty with $(\mathbf{g}, \mathbf{u}_n)/\tilde{\sigma}_n$ still exists. In application it is often found that, whereas the first few terms of the series for $\mathbf{f}$ decrease, the remaining ones tend to get larger and larger, although not necessarily monotonically. Truncation is again suggested, and this is frequently done by throwing out these increasing terms on the grounds that they are primarily evidence of the noise in the data.

This approach is often fairly successful, although it must be recognized that the truncation criterion may be somewhat arbitrary. There is no reason to believe that some of the terms rejected may not contain some valuable information. If a valid estimate of the data error is available, we may develop a more sophisticated criterion for truncation.

It is clear that this singular value decomposition device can have an advantage over others that we have discussed. Relatively rational "stopping rules" can be used to determine where the series for f should be truncated. At the same time, at least some "subjective behavior" is relatively harder to build into the solution function f. For instance, with the projection method of the previous section, violent oscillations in f can be avoided by projecting on a space that contains no highly oscillatory functions. With the present device, elimination of such oscillations may be more difficult.

As you doubtless have observed I have implicitly assumed that you have available software capable of providing singular vectors and singular values. Such programs are now quite common, and are frequently referred to as singular value decomposition schemes [18, 42]. If your equipment does not have this capability, I hope that the basic ideas will still be clear.

6.4.2 A Variant of the Method

Because singular functions (or vectors) are somewhat unfamiliar and relatively complicated to compute, we might try to avoid them by the following trick. Consider $g = Kf$. Operate on both sides with K^* to get

$K^*g = K^*Kf$. This new IFK seems to have several advantages. Because K^* is an integral operator, it will smooth the data g. Moreover K^*K is a symmetric operator and so is pleasant to deal with. I pursue this attack briefly.

Recall from Section 5.4 that the eigenvalues of K^*K are just σ_n^2, where the σ_n are the singular values of K. Moreover, since K^*K is symmetric, its singular functions are just its eigenfunctions ϕ_n. Hence the solution to the new IFK can be written (see Eq. (5.19))

$$f = \sum_{n=1}^{\infty} \frac{(K^*g, \phi_n)}{\sigma_n^2} \phi_n.$$

This equation reveals that my "trick" has probably gained us little. Although K^*g may be smoother than g, the eigenvalues σ_n^2 approach zero more rapidly than do the σ_n, and the effects seem to balance out. Indeed, it can be shown by reintroducing arguments concerning singular functions that *nothing* has been gained. I leave details to you if you are interested in them; you might also wish to study further the obvious connection between this method and least squares.

(Of course, if your computing capabilities allow the calculation of eigenvalues and vectors of symmetric matrices, but not of singular values and vectors of nonsymmetric matrices, the device described here may be useful.)

6.5 Parametrized Solution Function Method

The device that I shall now briefly describe is probably even older than the quadrature approach. Although it is intuitively quite obvious, the quadrature approach can be implemented effectively only if fairly large systems of linear algebraic equations can be solved. This, in turn, assumes the availability of a computer. The parametrized function method may require less computing. At the same time, it is quite ineffective, and often entirely misleading, unless we have good subjective knowledge concerning the solution of the IFK.

Implementation requires that we guess a solution *form* to within a few parameters. For instance, if we feel quite certain that the solution is a Gaussian, then we may try

$$f(y) = \alpha e^{-\beta(y-\gamma)^2},$$

where α, β, and γ are parameters that must be determined. This may be done by least squares fitting to the data. A total of three nonlinear algebraic equations for the unknown parameters α, β, and γ is obtained. Although the nonlinearity is unpleasant, the small number of equations can often be handled with relatively crude equipment.

It is clear that this device is meaningful only if the solution shape is quite certain. Moreover, not too many parameters should be used when the least squares equations are nonlinear. It is unfortunate that sometimes a fairly good least squares fit to the data can be obtained even though a very poor guess has been made of the shape of the solution function.

Finally, we notice that the method of Sections 6.3.1 and 6.3.2 is actually included in the current discussion. In that case, the restriction that the number of parameters (the series coefficients) be small can be removed because the least squares equations are linear. A combination of the two approaches is possible. For instance, a "correction" series might be added to the Gaussian

$$f(y) = \alpha e^{\beta(y-\gamma)^2} + \sum_{n=1}^{N} f_n \theta_n(y).$$

Here the θ_n are given and the f_n (as well as α, β, and γ) are to be calculated. Although the additional coefficients f_n appear in a linear fashion, the equations that determine the parameters remain nonlinear and the solution may be difficult for N at all large.

6.6 The Method of Regularization

6.6.1 Some Background and the Basic Scheme

The algorithm that I shall now discuss is currently one of the most popular and most successful. It was introduced about 30 years ago in this country by Phillips [41]. His work was extended a year later by Twomey [53]. At the same time Tikhonov, apparently unaware of the efforts of Phillips, published a paper on the subject in the USSR [50]. The method of *regularization* involves an idea whose "time had come." I shall explain the approach in a context somewhat different from that originally used by Phillips.

Suppose that we can guess a reasonable approximate solution f_0 to the IFK $g = Kf$. Write $F = f - f_0$. Then we anticipate that F is small.

Moreover F satisfies an IFK:

(6.8) $$g - Kf_0 = G = KF.$$

Now form the functional

$$\begin{aligned}\mathcal{F}(W,\gamma) &= \int_a^b dz \left[G(x) - \int_a^b K(x,y)W(y)dy \right]^2 + \gamma \int_a^b W^2(y)dy \\ &= \| G - KW \|_2^2 + \gamma \| W \|_2^2, \quad \gamma > 0.\end{aligned}$$

If $W = F$ the first term of this functional is zero. If W is small, the second term is small. We should like to *minimize* $\mathcal{F}$ by adroit choice of W, thus obtaining a function that "almost solves" Eq. (6.8) and that is at the same time small.

So far γ has not been specified. Suppose γ is itself small. Then the functional $\mathcal{F}$ puts the heavier weight on the first term; the minimizing function comes close to solving Eq. (6.8) but may not be especially small. If γ is large, the smallness of W is emphasized, and the minimizing function may not be a particularly good approximation to the solution of Eq. (6.8). Obviously, γ can be juggled as desired.

The minimization of the functional for fixed γ is a problem in the calculus of variations. Without further discussion I merely state that it can be shown that W, which I now write as W_γ to emphasize its dependence on the parameter, satisfies an integral equation of the *second kind*,

(6.9) $$\gamma W_\gamma = K^*G - K^*KW_\gamma$$

or, in more standard form,

$$W_\gamma = \frac{K^*G}{\gamma} - \frac{1}{\gamma} K^*KW_\gamma.$$

Observe that the parameter γ of Section 5.5 now appears as $(-1/\gamma)$, while the function referred to as g in that section is now K^*G/γ. Note, too, that K^*K is symmetric and recall that its eigenvalues are nonnegative (Section 5.4). If we write those eigenvalues as λ_n^2 and the corresponding eigenfunctions as ϕ_n, the solution to Eq. (6.9) can be given, using Eq. (5.18), in the form

$$W_\gamma(x) = \frac{K^*G}{\gamma} - \sum_{n=1}^\infty \frac{\lambda_n^2}{\gamma + \lambda_n^2} \left(\frac{K^*G}{\gamma}, \phi_n \right) \phi_n(x).$$

Recall that this solution is unique.

Before proceeding, let us remember that our goal is to resolve the problem $G = KF$. If this equation has a solution F, then $K^*G = K^*KF$, which is to say that K^*G is a function representable as $(K^*K)F$. Therefore (see Eq. (5.9) et seq.) it is possible to write K^*G in a series of the eigenfunctions of K^*K:

$$K^*G = \sum_{n=1}^{\infty} (K^*G, \phi_n)\phi_n(x).$$

Thus

$$W_\gamma(x) = \sum_{n=1}^{\infty} \left(\frac{K^*G}{\gamma}, \phi_n\right) \phi_n(x) - \sum_{n=1}^{\infty} \frac{\lambda_n^2}{\gamma + \lambda_n^2} \left(\frac{K^*G}{\gamma}, \phi_n\right) \phi_n(x)$$

(6.10) $$= \sum_{n=1}^{\infty} \frac{(K^*G, \phi_n)}{\gamma + \lambda_n^2} \phi_n(x).$$

According to our earlier discussion, this equation may be particularly interesting for small γ. We *formally* allow γ to approach zero through positive values

$$\lim_{\gamma \to 0^+} W_\gamma(x) = \sum_{n=1}^{\infty} \frac{(K^*G, \phi_n)}{\lambda_n^2} \phi_n(x).$$

We rewrite this using a basic property of K^*:

$$\lim_{\gamma \to 0^+} W_\gamma(x) = \sum_{n=1}^{\infty} \frac{(G, K\phi_n)}{\lambda_n^2} \phi_n(x).$$

Finally, we recall that in the notation used for singular functions (see Section 5.4),

$$\phi_n = v_n, K\phi_n = Kv_n = \sigma_n u_n, \lambda_n = \sigma_n.$$

Thus

$$\lim_{\gamma \to 0^+} W_\gamma(x) = \sum_{n=1}^{\infty} \frac{(G, u_n)}{\sigma_n} v_n(x).$$

But if $G = KF$ then F has the expansion (see Section 5.6)

$$F(x) = \sum_{n=1}^{\infty} \frac{(G, u_n)}{\sigma_n} v_n(x).$$

Thus we have the formal result

$$\lim_{\gamma \to 0^+} W_\gamma(x) = F(x).$$

This is not too surprising because Eq. (6.9) yields, again completely formally,

$$K^*G = K^*K \left(\lim_{\gamma \to 0^+} W_\gamma \right).$$

Before going further, let us discuss how these ideas may be put into practice. We consider the integral equation, Eq. (6.9). To solve it, we write an approximate matrix equation, using (probably) a quadrature scheme:

(6.9′) $$\mathbf{W}_\gamma = \frac{\mathbf{K}^*\mathbf{G}}{\gamma} - \frac{1}{\gamma}\mathbf{K}^*\mathbf{K}\mathbf{W}_\gamma.$$

We solve this equation for smaller and smaller positive values of γ and attempt to find the "numerical limit" of the $\mathbf{W}_\gamma$. That limit will be an approximation to F, the solution to $G = KF$.

It should be mentioned that the numerical solution of integral equations of the second kind is relatively easy [5, 7, 8, 15, 16]. However, the "numerical limit" mentioned in the previous paragraph is not a simple matter. In fact, Eq. (6.10) reveals that numerical difficulties are certain to occur because G contains errors inherited from the data g. When both γ and λ_n are small, the term $(K^*G, \phi_n)/(\gamma + \lambda_n^2)$ will almost certainly become ill behaved, and the numerical process must stop.

In practice, we usually compute W_γ for smaller and smaller values of γ. Calculations are halted when W_γ begins to exhibit unpleasant behavior. The "most acceptable" W_γ is then selected as the "regularized" solution of the IFK.

To obtain some idea of the relationship between γ and the error in G let us rewrite Eq. (6.10), recalling the relations among $\phi_n, v_n, u_n, \lambda_n$, and σ_n (see Section 5.4),

$$\begin{aligned} W_\gamma(x) &= \sum_{n=1}^\infty \frac{(K^*G, \phi_n)}{\gamma + \lambda_n^2} \phi_n(x) \\ &= \sum_{n=1}^\infty \frac{(G, K\phi_n)\phi_n(x)}{\gamma + \lambda_n^2} \\ &= \sum_{n=1}^\infty \frac{\lambda_n(G, u_n)v_n(x)}{\gamma + \lambda_n^2}. \end{aligned}$$

Suppose that G is in error (an error inherited from g). We replace G by $G_c + \epsilon$, where G_c denotes the correct G, and note that

(6.11)
$$\sum_{n=1}^{\infty} \frac{\lambda_n(G_c + \epsilon, u_n)v_n(x)}{\gamma + \lambda_n^2} = \sum_{n=1}^{\infty} \frac{\lambda_n(G_c, u_n)v_n(x)}{\gamma + \lambda_n^2} + \sum_{n=1}^{\infty} \frac{\lambda_n(\epsilon, u_n)v_n(x)}{\gamma + \lambda_n^2}.$$

Clearly we need information about the error term

$$E(x) = \sum_{n=1}^{\infty} \frac{\lambda_n}{\gamma + \lambda_n^2}(\epsilon, u_n)v_n(x).$$

Now, for fixed γ, it is easy to see that $\lambda_n/(\gamma + \lambda_n^2)$ never exceeds $1/2\sqrt{\gamma}$, suggesting a quantitative relationship between γ and ϵ. With some knowledge of orthogonal series we may demonstrate that

$$\| E \|_2 \leq \frac{1}{2\sqrt{\gamma}} \| \epsilon \|_2 .$$

This suggests that a "safe" choice for γ may well be $\gamma > \| \epsilon \|_2^2 /4$. Indeed, such a value as an initial choice for γ is reasonable. You should not expect that it is automatically the "best" choice.

Notice that the presence of γ in the denominator of the other series on the right of Eq. (6.11) also distorts the solution. However, we expect γ to be small. For those n sufficiently large that γ affects the denominator significantly, the numerator is small, and the distortion thus is not great. Again, a complicated analysis justifying these rather vague remarks is possible, but would carry us far afield.

In the description of the regularization procedure, little use has been made of the functional $\mathcal{F}$. The integral equation, Eq. (6.9), has been useful in part because I have been able to demonstrate some of the ideas developed in this primer. In practice, direct minimization methods are often used. These have the advantage that constraints (for example, positivity) may be imposed on the solution-function relatively easily. This cannot be done using our approach. It should be remarked that imposition of constraints greatly complicates the mathematical analysis.

6.6.2 Some Numerical Examples

I shall discuss two numerical examples using the regularization procedure described in detail in the previous subsection. Others will be found in the problems.

Example 1

It seems desirable to return to the example of Section 6.2.2 to determine if regularization really achieves any improvement. Two trial functions f_0 are used, $f_0(y) = 0$ and $f_0(y) = 0.5$. The first suggests no prior knowledge of the solution. The second might be construed as a new attempt after some knowledge of $f(y)$ has been obtained. In all cases **K** was a 101 × 101 square matrix. Results are exhibited in Tables 6.4a,b and 6.5a,b. These were obtained by direct solution of Eq. (6.9′).

First, it is clear that the improvement over straight quadrature is quite remarkable. It is somewhat surprising that the "complete ignorance" guess $f_0 = 0$ is not significantly worse than the guess $f_0 = 0.5$.

No attempt was made to find the exact 2-norm of the error ϵ, but the magnitude of the better values of γ is about as expected. No effort was made to compute the best γ in any instance.

Of course, the values of ρ are somewhat misleading. They can only be computed because the true value of the solution is known. In a realistic situation this knowledge would not be available.

In practice, it is customary to plot the functions f for various values of γ and then select that one which "looks" best, taking into account any information about the solution f provided by the origin of the problem. Clearly this approach is purely subjective. By now you should be prepared for this kind of thing.

In Figs. 6.5 and 6.6 we give plots of f for $\bar{\epsilon} = 10^{-2}$ and $f_0 = 0$ using $\gamma = 10^{-4}$ and $\gamma = 10^{-3}$. Try to forget that the exact solution to the problem without data error is simply $f(y) = y$. Which figure would *you* choose?

Example 2.

Let us examine a different—and as yet undiscussed—kernel namely,

$$K(x,y) = ye^{-xy^2}, \qquad 0 \leq x, y \leq 1.$$

Observe that this K is not symmetric. (The kernel in Example 1 is symmetric.) Moreover, this K is very smooth. In the light of the remarks

Table 6.4a.
Results for Example 1, $f_0 = 0.0$
$\bar{\epsilon} = 10^{-4}$

γ	ρ
10^{-5}	$0.407(0)$
10^{-4}	$0.881(-1)$
10^{-3}	$0.529(-1)$
10^{-2}	$0.943(-1)$

Table 6.4b.
Results for Example 1, $f_0 = 0.0$
$\bar{\epsilon} = 10^{-2}$

γ	ρ
10^{-5}	$0.407(0)$
10^{-4}	$0.100(0)$
10^{-3}	$0.705(-1)$
10^{-2}	$0.123(0)$

Table 6.5a.
Results for Example 1, $f_0 = 0.5$
$\bar{\epsilon} = 10^{-4}$

γ	ρ
10^{-6}	$0.349(-1)$
10^{-5}	$0.344(-1)$
10^{-4}	$0.390(-1)$
10^{-3}	$0.497(-1)$

Table 6.5b.
Results for Example 1, $f_0 = 0.5$
$\bar{\epsilon} = 10^{-2}$

γ	ρ
10^{-5}	$0.407(0)$
10^{-4}	$0.881(-1)$
10^{-3}	$0.529(-1)$
10^{-2}	$0.943(-1)$

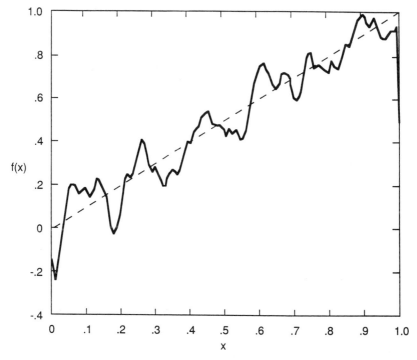

Figure 6.5. Computed f for Example 1, $\gamma = 10^{-4}$, $\bar{\varepsilon} = 10^{-2}$, $f_0 = 0.0$.

of Section 6.2.5 we anticipate increased difficulty. We try to solve

$$g(x) = \int_0^1 y e^{-xy^2} f(y) dy,$$

where

$$g(x) = \begin{cases} \frac{1}{2x}(1 - e^{-x}), & x > 0, \\ \frac{1}{2}, & x = 0. \end{cases}$$

It is easy to see that $f(y) \equiv 1$ is the solution.

We proceed as in Example 1 first using $\gamma = 0$ (that is, straightforward quadrature). Here there is total failure. The matrix $\mathbf{K}$ (still 101×101) is numerically singular, even on a 14-digit machine. The smoothness of $K(x,y)$ has evidenced itself.

This difficulty is eliminated by introducing $\gamma > 0$. We make the guess $f_0 = 0.75$ and obtain the results shown in Tables 6.6a,b.

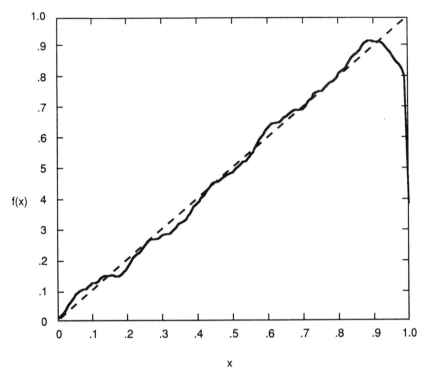

Figure 6.6. *Computed f for Example 1, $\gamma = 10^{-3}, \bar{\epsilon} = 10^{-2}, f_0 = 0.0$.*

The results are inferior to those for Example 1 with $f_0 = 0.5$. The graphs of the computed F's are even more disappointing. We present just one for the best case when $\epsilon = 10^{-4}$, Fig. 6.7. (Observe that $F(x) \equiv 0.25$).

It may seem especially distressing that the computed F is quite wrong at $y = 0$. Does this demonstrate a flaw in the theory? In a sense the answer is yes. For a partial explanation we must observe the footnote in Section 5.3 where it is remarked that the various series expansions only converge as expected when an appropriate interpretation is used. Here is a case in which ordinary pointwise convergence fails. (See [24] for some clarification.)

Indeed, consider the expansion of Section 6.6.1.

$$W_\gamma(x) = \sum_{n=1}^{\infty} \frac{\lambda_n (G, u_n) v_n(x)}{\gamma + \lambda_n^2}.$$

Table 6.6a.
Results for Example 2, $f_0 = 0.75$
$\bar{\epsilon} = 10^{-4}$

γ	ρ
10^{0}	0.132(0)
10^{-8}	0.830(−1)
10^{-7}	0.778(−1)
10^{-6}	0.800(−1)
10^{-5}	0.884(−1)

Table 6.6b.
Results for Example 2, $f_0 = 0.75$
$\bar{\epsilon} = 10^{-2}$

γ	ρ
10^{-7}	0.274(0)
10^{-6}	0.130(0)
10^{-5}	0.802(−1)
10^{-4}	0.899(−1)
10^{-3}	0.958(−1)

Now, by definition (see Section 5.4),

$$\sigma_n v_n(x) = \int_0^1 x e^{-x^2 y} u_n(y) dy$$

Clearly $v_n(0) = 0$, $n = 1, 2, \cdots$. Thus for any γ, $W_\gamma(0) = 0$. Hence, no solution computed by this regularization scheme can have the correct value at $x = 0$, unless we (cleverly) select an f_0 such that $f_0(0) = 1$. Such a choice would imply very unusual insight.

Nor can it be argued that direct minimization of the functional $\mathcal{F}$ might avoid the difficulty. Solution of the integral equation, Eq. (6.9), is equivalent to such a direct approach.

These remarks suggest that some information about the success or failure of the regularization method (and some of its variants) may be obtained by study of the singular values and functions of the kernel of the original integral equation. That is precisely the case, as Problems 10 and 11 serve to demonstrate.

The problems we have considered are not of the form originally studied by Phillips [41]. He did not guess a solution f_0. Rather he imposed

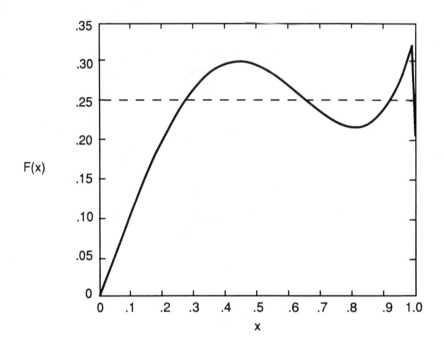

Figure 6.7. *Computed F for Example 2, $\gamma = 10^{-7}, \bar{\epsilon} = 10^{-4}, f_0 = 0.75$.*

the restriction that the solution to his problem should not oscillate too rapidly, and used as a measure of the total oscillation of f, the expression

$$\int_a^b |f''(y)|^2 \, dy = \| f'' \|_2^2 .$$

The functional $\mathcal{F}$ is no longer appropriate. It is replaced by

$$\mathcal{H}(W, \gamma) = \| g - KW \|_2^2 + \gamma \| W'' \|_2^2 .$$

The goal is now to minimize this functional for various values of the parameter γ. If we use the arguments of the calculus of variations, we discover that Eq. (6.9) is replaced by an integrodifferential equation. Equations of this kind are less well understood than many we have been studying. However, direct minimization of $\mathcal{H}$ can be carried out.

It is probably clear to you that many variations of this regularization approach are possible. For example, one might require that the derivative of f should be small, or that it be approximately equal to some prescribed function. The possibilities are endless, and many have been used in practice.

6.6.3 The Selection of the Parameter, γ

You may be somewhat distressed at the comment in Example 1 that the regularization parameter γ is selected on the basis of visual inspection of the computed solution. Although in practice this is often the approach used we may ask, "Isn't there a somewhat less subjective way?"

You might guess that the choice of γ is determined just by K and by the statistical distribution of the error ϵ in the data. Unfortunately, this is not true, as can be demonstrated by numerical example. (See, e.g., [31].) The data, g, itself plays a vital role in the determination of the optimal γ. There are many analytical approaches to this problem. It is not appropriate to go into detail in this *Primer*. If you are interested, see, for instance [19, 23, 25, 39, 51, 52, 54].

6.6.4 A Generalization of the Concept of Regularization

The devise emphasized in this section is often referred to as Tikhonov regularization. Perhaps the functional $\mathcal{F}$ (see Section 6.6.1) best indicates what it does. It makes a compromise between ill posedness and accuracy.

Actually we have done this sort of compromising before. In Section 6.2 we selected the "best" number of quadrature points (hence the "best" mesh size). In Section 6.4 we truncated at a favorable value of n the number of terms retained in the series expansion. Similar ideas will be employed in material which follows.

If we consider all such compromises as "regularizations," the concept is considerably extended. Formal work along these lines has been done and some of it is nicely summarized in [33]. (See also Problem 16.)

6.7 Stochastic Approaches

The fact that all data in our problems are expected to be in error or contain "noise" suggests that some sort of statistical or stochastic investigation be made. This approach has been taken by numerous investigators; see, for example, [21, 44]. The basic assumption that our data is sparse makes this stochastic method less appealing than it would be if the data were ample because the assignment of meaningful statistics may be somewhat unrealistic. However, this method should be discussed briefly, if only to show its relationship to some other schemes. We follow the approach of Franklin [21].

The standard IFK, $g = Kf$, can be rewritten as

$$g = -\epsilon + Kf,$$

where ϵ now represents the noise in the data g. Somewhat more generally we can examine

$$\xi_3 = \xi_2 + K\xi_1,$$

where the ξ's are random variables. It is convenient to assume that ξ_1 and ξ_2 have mean zero. If this is not the case, we introduce

$$\zeta_i = \xi_i - \bar{\xi}_i, \qquad i = 1, 2,$$

where $\bar{\xi}_i$ is the mean of ξ_i. After a bit of algebra we find the new equation

(6.12) $$\zeta_3 = \zeta_2 + K\zeta_1,$$

where ζ_1 and ζ_2 have mean zero.

Next, we assume knowledge of the expected values of the random variables

$$E\{\zeta_i(t), \zeta_j(t')\} = \rho_{ij}(t, t')$$

and define cross-correlation and auto-correlation operators

(6.13) $$R_{ij}(h_1, h_2) = \int_a^b \int_a^b \rho_{ij}(t, t') h_1(t) h_2(t') dt dt'.$$

In many problems it is reasonable to suppose there is no correlation between signal and data noise so that $R_{12} = R_{21} = 0$. In this case it can be shown that

$$R_{33} = K R_{11} K^* + R_{22}.$$

Under various other assumptions, which I choose not to present in detail, it can also be demonstrated that a best linear estimate to f, the solution to $g = Kf$, is given by

(6.14) $$f = R_{11} K^* (K R_{11} K^* + R_{22})^{-1} g.$$

This slightly forbidding expression is illuminated by recalling that the translations used in obtaining zero means are reminiscent of the translations used in the preceding section. There we eventually encountered Eq. (6.9), which leads to a "regularized" solution of $g = Kf$. If we use the inverse of K^*, Eq. (6.9) can be rewritten

$$\gamma(K^*)^{-1}W_\gamma = G - KW_\gamma$$

so that

$$\begin{aligned}
W_\gamma &= (K + \gamma(K^*)^{-1})^{-1}G \\
&= [(KK^* + \gamma I)(K^*)^{-1}]^{-1}G \\
&= K^*(KK^* + \gamma I)^{-1}G \\
&= \eta K^*(K\eta K^* + \gamma \eta I)^{-1}G,
\end{aligned}$$

where η is an arbitrary positive constant.[2] This equation is very similar to Eq. (6.14). In fact, there is complete agreement if $R_{11} = \eta I$ and $R_{22} = \gamma \eta I$. This is precisely the case of white noise, obtained when ρ_{11} in Eq. (6.13) and the corresponding ρ_{22} for R_{22} are taken as delta functions times η and $\eta\gamma$, respectively.

Thus a close relationship is seen between this stochastic method and regularizaton of the sort studied in some detail in the previous section. Other regularization devices can lead to other operators R_{11} and R_{22}. A further study clarifies the relationship between γ and the error expected in g. For more information, see [38, 44].

The above discussion should not be used as an argument that the stochastic approach can be totally ignored in favor of regularization. Both methods and viewpoints have their advantages. It is good to realize that they are actually connected. Personal taste may decide the approach used.

I point out that the method of generalized cross validation of Wahba clearly combines the idea of regularization with certain statistical considerations. Explicit use of the operators R_{ij} is avoided (see, for example, [54]).

---

[2] Although the K's are integral operators, they may be formally manipulated in the same way as matrices. The symbol I denotes the identity operator.

6.8 Iterative Methods

In this day of iterative approaches to so many problems it may seem strange that I have not mentioned iteration until this point. I have preferred to postpone discussion until meaningful remarks could be made concerning the relationship between iteration and other schemes.

Consider, for example, the regularization method discussed in detail in Section 6.6. There we guessed an f_0 thought to be close to f, and then resolved the reduced problem by finding W_γ. If we define

$$f_1 = f_0 + W_\gamma,$$

f_1 should be a still better estimate of the solution to $g = Kf$. The process may obviously be repeated.

This rather simple idea may be also be applied with variations to the stochastic approach. Some success has been achieved.

A class of "pure" iteration schemes also exists of which the Landweber method [34] is representative:

$$(6.15) \qquad f_k = f_{k-1} + \alpha K^*(g - Kf_{k-1}), \qquad k = 1, 2, \cdots,.$$

Here f_0 is again a "good guess" and α is a suitable constant. Under appropriate circumstances, $f_k \to f$. Practical implementation of Eq. (6.15) is sometimes possible, but usually the successive iterates begin to become ill-behaved, and numerical divergence occurs. The f_k frequently develop the oscillatory behavior so characteristic of approximate solutions to IFKs.

Diaz and Metcalf [17] have shown that a wide class of iterative methods of the Landweber form is really equivalent to the singular function expansion of Section 6.4. These devices hence suffer from exactly the same difficulties as that expansion. They have the advantage that the singular functions need not be explicitly calculated, and the disadvantage of concealing to some extent what is actually going on. It can be shown that such iterative methods converge if and only if the corresponding expansion converges, and that seldom happens numerically.

We observed earlier that the singular function expansion solution may simply be more or less arbitrarily truncated when difficulties set in. Is there a similar ad hoc device for stopping the iteration? Obviously, we can terminate when the iterates seem no longer to be converging.

A more sophisticated approach has been developed by Strand [48] who introduced a modified Landweber algorithm:

$$f_k = f_{k-1} + DK^*(g - Kf_{k-1}).$$

Here D is a filtering operator designed to attenuate the iteration so as to basically cut out all but the first few terms of the singular function expansion (which, of course, is not explicitly obtained). Although quite neat and automatic, this filtering scheme is obviously highly subjective. The problem solver must choose the filter. Whether we wish to make this choice or prefer simply to stop the iteration "when the going gets tough" may again be a matter of taste.

6.9 Summary

In this chapter I have presented a few methods for resolving integral equations of the first kind. An attempt has been made to select relatively common or promising ones. None has been described in great detail. Many devices have not been mentioned, as a glance at the voluminous research literature will reveal. Many of the algorithms not covered have relatively limited use, often being applicable successfully only to a small class of problems. Others, while important, are variants on the schemes presented here, or combinations of those schemes. There are quite likely numerous schemes of which I am unaware. For such cases I apologize.

Throughout the treatment, the necessity of making subjective judgments, using insight, intuition, and educated guesses has been clear. Personal taste certainly plays a role. One individual may choose to use Tikhonov regularization whereas the next may feel much more comfortable with a stochastic approach. It is important to know what methods are available and what their basic advantages and disadvantages are and to be willing to experiment with new ones and vary old ones. Above all, we should constantly recognize that wishful thinking does not replace insight and judgment. It often happens that we can almost "will" a "solution."

Problems VI

1. Make a numerical study of the results found for the problem

$$g(x) = \int_0^1 e^{-\alpha|x-y|} f(y) dy$$

using different quadrature schemes of your choice. Continue to use the errors $\bar{\epsilon} = 0$, 10^{-4}, and 10^{-2}.

2. Using available software, compute the condition numbers for the matrices arising in Problem 1 as well as for the trapezoidal scheme of the text. Compare these numbers with the errors in g and f.

3. Construct the functions $\tilde{\alpha}_i$ and $\tilde{\beta}_j$ of Section 6.2.7.

4. Repeat the numerical experiments of Section 6.2.7 using

   (a) $K_n(x,y) = \sum_{j=0}^{n} \frac{(-1)^j (xy)^{2j}}{(2j)!}, 0 \leq x, y \leq 1,$

   $K(x,y) = \cos xy$

   (b) $K_n(x,y) = \sum_{j=0}^{n} \frac{(-1)^j (xy^2)^{j/3}}{j!}, 0 \leq x, y \leq 1,$

   $K(x,y) = e^{-x^{1/3} y^{2/3}}.$

5. Consider the problem of Section 6.2.4, but now use fewer quadrature points than data points (for example, 21 values of x_i and 6 values of y_i). Solve this resulting least squares problems. Compare the results with those obtained in the text.

6. Discuss the computational difficulties that may arise in implementing the ideas of Sections 6.3.1 and 6.3.2 assuming that the double integrals defining k_{ij} cannot be evaluated analytically. Observe that an attempt to overcome these difficulties might lead you to unwisely select θ's and ψ's which are not really appropriate to the problem under consideration.

   If you have the computing power available try the Galerkin approach on some of the problems we have examined using θ and ψ sets of your choosing.

7. Consider the N by N matrix problem $\mathbf{g} = \mathbf{Kf}$. Prove that if this problem has a unique solution it is given by

$$\mathbf{f} = \sum_{n=1}^{N} \frac{(\mathbf{g}, \mathbf{u}_n) \mathbf{v}_n}{\tilde{\sigma}_n},$$

where the $\mathbf{u}_n$ and $\mathbf{v}_n$ are the singular vectors of $\mathbf{K}$ and the $\tilde{\sigma}_n$ are the singular values.

Resolving Integral Equations of the First Kind   109

8. In Problem 7 (and in the text material) the possibility arises that some $\tilde{\sigma}_n$ might be zero. Recall that $\tilde{\sigma}_n^2$ is an eigenvalue of $\mathbf{K}^*\mathbf{K}$ and that the determinant of this matrix has value $\Pi_{n=1}^{N}\tilde{\sigma}_n^2$. What does this lead you to conclude about the invertability of $\mathbf{K}$?

9. Verify by straightforward substitution that $f(y) = y$ solves

$$\int_0^1 f(y) \cos\left(\frac{y^2}{x+1}\right) dy = \frac{x+1}{2} \sin\left(\frac{1}{x+1}\right), \qquad 0 \le x, y \le 1.$$

Now guess a solution of the form

$$f(y) = \alpha \sin \beta y.$$

Using the method of Section 6.5 and ten datapoints x_i, use the method of least squares to determine α and β. Determine the goodness of fit to both the data function and the solution function. Repeat the calculation using the guess

$$f(y) = \tilde{\alpha} \sin \tilde{\beta} y + \tilde{\gamma}.$$

10. Consider the IFK

$$g(x) = \int_0^1 y e^{-xy^2} f(y) dy,$$

where, with $y_2 > y_1$,

$$g(x) = \begin{cases} \frac{1}{2x}[e^{-xy_1^2} - e^{-xy_2^2}], & x > 0, \\ \frac{1}{2}(y_2 - y_1), & x = 0. \end{cases}$$

Show by direct substitution that

$$f(y) = \begin{cases} 1, & y_1 < y < y_2, \\ 0, & \text{all other } y. \end{cases}$$

It should not be anticipated that any standard regularization method will yield good results because such methods represent $f(y)$ as (finite) sums of smooth functions. Nevertheless, try the problem for the following three cases (use $N = 101$):

(a) $y_1 = 0.05, y_2 = 0.15$;

(b) $y_1 = 0.5, y_2 = 0.6$;

(c) $y_1 = 0.8, y_2 = 0.9$.

Which result is best? Worst?

11. Compute the singular values and singular functions for the kernel of the previous problem. (You will only be able to obtain a few because the σ_n get small quickly.) In the light of your results are your findings in Problem 10 surprising?

    Explain how knowledge of the σ_n, v_n, and u_n may be helpful in more general situations. Could they be useful in the design of physical experiments?

12. Again consider
$$g(x) = \int_0^1 y e^{-xy^2} f(y) dy.$$

Set
$$y = \sqrt{z}$$

to get
$$g(x) = \frac{1}{2} \int_0^1 e^{-xz} f(\sqrt{z}) dz.$$

Define
$$f(\sqrt{z}) = \hat{F}(z)$$

so that
$$g(x) = \frac{1}{2} \int_0^1 e^{-xz} \hat{F}(z) dz.$$

Observe that the kernel of this equation is now symmetric. Redo Problems 10 and 11. Any improvements? (Note: such simplifications are only infrequently possible.)

13. If you have a good knowledge of orthogonal series, show that (see Section 6.6.1)
$$\| E \|_2 \leq \frac{1}{2\sqrt{\gamma}} \| \epsilon \|_2 .$$

14. If you are familiar with the calculus of variations, obtain Eq. (6.9). This is actually the Euler equation for the minimization of $\mathcal{F}$.

15. Write out explicitly the first few terms in Eq. (6.15) obtaining expressions which contain only f_0 and g. Obtain a general expression. Under what conditions does convergence seem to occur?

16. Attempt a numerical resolution of the equation of Problem 9 using the iteration scheme provided by Eq. (6.15). Experiment with different values of the parameter α.

17. Rewrite Eq. (6.7) as

$$f(y) = \sum_{n=1}^{\infty} \Lambda_n \frac{(g, u_n)}{\sigma_n} v_n(y),$$

where, clearly, $\Lambda_n \equiv 1$.

(a) Select the Λ_n's so that the series truncates at $n = N$.
(b) Select the Λ_n's so as to obtain Tikhonov regularization.
(c) Consider other possible choices for Λ_n, bearing in mind that you are always compromising between ill posedness and accuracy. (See [33].)

CHAPTER 7

Some Important Miscellany

7.1 Introduction

During the preparation of this book, I have often been tempted to digress and make side comments. For the most part (I hope!) this desire has been resisted in order not to interrupt the more or less orderly flow of concepts and ideas. However, some of the matters not discussed are relatively important. I present here a potpourri of miscellaneous remarks, concepts, and ideas.

7.2 About Those Ground Rules

Why have I insisted the δ-function not be used? That a and b be finite? The functions well behaved? Real?

Had I assumed that you were familiar with the Lebesgue integral I could have relaxed some of these constraints. Most of the results stated are rigorously correct if the requirement is made that $K^2(x,y)$ is integrable, that $f^2(y)$ is integrable, and that convergence is always interpreted in the L_2 sense. (That is, in the 2-norm sense. See Appendix C). In this context $a = -\infty$ and/or $b = +\infty$ are completely acceptable.

One must be cautious, however. Kernels which are perfectly satisfactory for a and b finite may not be square integrable when $b = \infty$ (or $a = -\infty$). For example, consider $K(x,y) = e^{-\alpha|x-y|}$, $a = 0$, $b = \infty$. Observe that

$$\int_0^\infty \int_0^\infty K^2(x,y)dxdy = \int_0^\infty \int_0^\infty e^{-2\alpha|x-y|}dxdy$$
$$= \int_0^\infty \int_{-y}^\infty e^{-2\alpha|u|}dudy$$
$$\geq \int_0^\infty dy \int_0^\infty e^{-2\alpha u}du = \frac{1}{2\alpha}\int_0^\infty dy = +\infty$$

(see Problem 1).

The δ-function is not square integrable in the Lebesgue sense. Many of the results I have discussed and used simply become incorrect for the δ-function.

The condition that all quantities be real is somewhat arbitrary. Much of our work remains correct in the complex framework provided norms, inner products, etc., are defined in accordance with Appendix B. I have chosen to avoid the relatively minor complications introduced by complex functions simply for ease of exposition.

7.3 A Final Word About Codes and Quadrature Schemes

I have assiduously avoided detailed discussions of the specifics of quadrature algorithms. No codes have been presented, although it has been indicated that certain software packages (such as the singular value decomposition) are valuable. My reasons are several.

It is certainly true that a quadrature method fitted to the problem at hand (for example, chosen because of the properties of $K(x,y)$) may often considerably improve numerical results or decrease computing time. Many research papers emphasize these considerations. However, when this is done the effectiveness of the basic concept—regularization, parametrization, etc.—can easily be obscured. I have chosen to keep the numerics as simple as possible. You are urged to carry out experiments with different integration schemes, placement of quadrature points, etc. You will find your level of success very problem dependent.

It is because of such considerations that no codes are presented. I have yet to see a universally satisfactory code for the resolution of IFKs. The discussion in this book should make it clear that it is unlikely that one can exist. You are invited to write your own software for classes of problems of interest to you. Just don't be surprised when it fails completely on a problem of a slightly different nature.

7.4 Finding Generalized Moments of Solutions

Sometimes the physical origins of a problem suggest that detailed knowledge of the solution $f(y)$ is not needed. Perhaps its integral, or one or two moments suffice. Is it possible to obtain such information without first computing f?

The answer is sometimes yes. The basic idea is found in so many works and in so many guises that it seems not even to have a standard name although the term "mollifier" is sometimes associated with aspects of the method. The following exposition is due to Golberg [22].

As usual, suppose f satisfies

$$g = Kf.$$

However, assume we really need only

(7.1) $$I = \int_a^b h(y)f(y)dy,\text{[1]}$$

where h is a given function, for instance, $h(y) \equiv 1$, $h(y) = y$, or $h(y) = y^2$.

Suppose it is possible to solve for $\theta(y)$ the new IFK

(7.2) $$h = K^*\theta.$$

(Observe that it may be that no solution to Eq. (7.2) exists.)

Rewrite (7.1), using basic properties of K^*:

(7.3) $$\begin{aligned} I &= (h, f) = (K^*\theta, f) = (\theta, Kf) \\ &= (\theta, g). \end{aligned}$$

Thus the problem of computing I is reduced to the trivial calculation of an inner product.

At first there seems to be no advantage. We have still had to resolve an IFK, namely, Eq. (7.2). But h is presumably known exactly. Thus there is no data error in Eq. (7.2), a considerable gain. And if g is in error, the error in I can be easily calculated (or estimated) from Eq. (7.3).

Also, suppose there are several data functions $g_1, g_2, \cdots, g_p$ to be studied. A single solution for θ followed by p inner product calculations yield all the desired I values.

---

[1] I is called a linear functional of f.

Of course, if Eq. (7.2) has no solution the method fails. But if all problems are discretized this difficulty is avoided. (There are many interesting mathematical questions lurking here. We cannot pursue them.)

Finally, a shrewd choice of h would seem to be the forbidden

$$h(y) = \delta(y - \hat{x});$$

then (see Eq. (7.1))

$$I = \int_a^b \delta(y - \hat{x}) f(y) dy = f(\hat{x}),$$

and the value of f at $\hat{x}$ has been found. But Eq. (7.2) is now insoluble.

This problem can be partially overcome by choosing h to be a function very sharply peaked at $\hat{x}$, falling rapidly away from $y = \hat{x}$. (Often h is chosen as a Gaussian.) Assuming h is normalized, we find

$$I \simeq f(\hat{x}).$$

This device has been employed by many researchers with considerable success. (See [4, 6, 35].)

7.5 IFKs in More Variables

Throughout this work I have dealt with integral equations which involve data dependent on a single variable, unknown functions dependent on just one variable, and kernels with two variables. You have perhaps encountered equations of the form

(7.4) $$g(x_1, x_2) = \int_a^b \int_a^b K(x_1, x_2; y_1, y_2) f(y_1, y_2) dy_1 dy_2$$

or with even more x's and y's. Can we say anything about those?

At the analytical level the concepts of singular values, singular functions, series expansions of solutions, etc., all carry over. In fact, problems even more general than Eq. (7.4) can be handled. Function analytical methods allow the study of general operator equations of the form

$$Tf = g$$

provided T is a "sufficiently smooth" linear operator [32, 33, 49].

However, approximate methods introduce a basic difficulty. If Eq. (7.4) is discretized in almost any fashion we have discussed, coefficients k_{i_1,i_2,j_1,j_2} arise. We no longer have a square or rectangular matrix. Devices to overcome this difficulty are known, but we cannot discuss such matters here.

7.6 Errors in the Kernel

We have assumed throughout that $K(x,y)$ is known exactly, and have given physical arguments to make such an assumption plausible. But it must be admitted that IFKs do arise with both K and g in error.

Such matters may be analyzed by writing $K(x,y) = K_T(x,y) + \delta K$, where K_T is the correct kernel, which we suppose is somehow obtainable, and δK is an estimate of the error. Clearly, detailed analyses can become somewhat messy. Perhaps the most important result is that if $\delta \mathbf{f}$ and $\delta \mathbf{K}$ are both small then Eq. (6.5) must be modified by simply adding a term involving $\| \delta \mathbf{K} \|_2$. The condition number C is computed from $\mathbf{K}_T$. (See [20, 47].)

7.7 Summary

In this chapter I have mentioned some miscellaneous matters which, though important, have seemed not quite to belong in earlier discussions. During the preparation of this chapter other matters have come to mind. However, more and more of those are highly specialized, often applying just to a narrow class of problems. I have tried throughout to avoid all but very general concepts and methods. If you are especially interested in seismology, tomography, radiative transfer, or some other particular field, I refer you to the literature. Perhaps some of the ideas gleaned from this book will aid you in better understanding the material found there.

Problems VII

1. Assume $\int_0^\infty K^2(t)dt = A > 0$ exists. Prove that

$$\int_0^\alpha \int_0^\infty K^2(|x-y|)dxdy$$

does not exist.

2. Consider the kernel $K(x,y) = \cos x^2 y$, $0 \le x, y \le 1$ and the equation $Kf = g$. Suppose we are interested only in

$$I = \int_0^1 h(y) f(y) dy,$$

where $h(y) = \sin y / y$. Show that $K^*\theta = \sin x/x$ has the solution $\theta(y) = y/2$.

Let $g(x) = \sin x^2 / x^2$. Verify that $f(y) = 1$. Show explicitly that $(h, f) = (g, \theta)$.

Pick a new function $\tilde{f}$ such that you can carry out the integration $K\tilde{f}$ analytically. Call the result $\tilde{g}$. Verify that $(\tilde{g}, \theta) = (\tilde{f}, h)$.

Now pick a more or less arbitrary $\tilde{g}$. Find $\tilde{f}$ computationally and repeat the above procedures numerically.

3. (a) Show that the result of Section 7.4 generalizes easily to matrices and vectors.

   (b) Suppose that the matrix equation **K f = g**, where **K** is square, must be solved for many different vectors **g** but that only the ith component of **f** is of any interest. Show how Problem 3a may be employed to do this.

   (c) Implement 3b for a specific 5×5 matrix **K**.

4. Consider the equation

$$g(x) = \int_x^1 \frac{f(y)}{\sqrt{y-x}} dy = Kf, \qquad 0 \le x \le 1.$$

   (a) Find the operator K^*.

   (b) Using Section 7.4 show how to compute

$$\int_0^1 x^n g(x) dx$$

   for any positive integer n.

CHAPTER 8

Epilogue

I hope that this brief and informal discussion of integral equations of the first kind may help those actually encountering such relatively unfamiliar equations in their work. It must also be fully admitted that matters have been left in quite an unsatisfactory state. No neatly tied package of solution methods has been produced. None can be.

You may, in fact, feel highly uneasy, especially when you consider carefully how great a role subjective thinking plays in resolving IFKs. Is there not some way of getting around the problem?

The obvious answer to this question is "Yes. Design experiments that do not lead to such equations." Such advice ignores the fact that many common experimental techniques give rise to IFKs and those techniques are not easily replaced. There may be no known alternatives, or the alternatives may simply be too difficult, too expensive, or require too much equipment, space, or manpower.

Given these considerations, can anything be done? Again the answer is yes. When any experiment gives rise to an IFK, an analysis of that IFK should be made *before* the experiment is actually performed. We have seen that some problems are more easily resolved than others. (For instance, the "flatter" the kernel $K(x,y)$, the less satisfactory it is.) Mathematical and computational investigation may reveal ways in which relatively simple changes may be made in an intended experiment to make its analysis significantly easier. This kind of study often requires little effort in terms of money, time, and manpower when compared to the investment in the experiment itself. If such a study reveals that the experiment cannot possibly generate useful information (to look at the most negative outcome) or that a few modifications can change a marginal experiment to a genuinely useful one (a more positive outlook), the analysis is well worthwhile. Tons of useless data stored in the archives may help to justify this approach.

References

1. ABRAMOWITZ, M., AND STEGUN, I. A., *Handbook of Mathematical Functions*, National Bureau of Standards, U. S. Government Printing Office, Washington, DC, 1964.

2. AKHIEZER, N. I., AND GLAZMAN, I. M., *Theory of Linear Operators in Hilbert Space*, translated by M. Nestell, Frederick Ungar, New York, 1961.

3. ALLEN, R. C., Jr., BOLAND, W. R., FABER, V., AND WING, G. M., *Singular values and condition numbers of Galerkin matrices arising from linear integral equations of the first kind*, J. Math. Anal. Appl., 109 (1985), pp. 564–590.

4. ANDERSSEN, R. S., *On the use of linear functionals for Abel-type integral equations in application*, in *The Application and Numerical Solution of Integral Equations* (Anderssen, R. S., et al., Eds.), Sijthoff and Noordhoff, Alphen aan den Rijn, 1980.

5. ATKINSON, K., *A Survey of Numerical Methods for the Solution of Fredholm Integral Equations of the Second Kind*, Society for Industrial and Applied Mathematics, Philadelphia, 1976.

6. BACKUS, G., AND GILBERT, F., *Uniqueness in the inversion of inaccurate gross earth data*, J. Phil. Trans. Roy. Soc. London Ser. A 266 (1970), pp. 123–192.

7. BAKER, C. T. H., *The Numerical Treatment of Integral Equations*, Clarendon Press, Oxford, 1977.

8. BAKER, C. T. H., and MILLER, G. F., *Treatment of Integral Equations by Numerical Methods*, Academic Press, London, 1982.

9. BANACH, S., *Théorie des Opérations Linéaires*, Hafner, New York, 1932.

10. CARTER, D. S., PIMBLEY, G., AND WING, G. M., *On the unique solution for the density function of PHERMEX*, Technical Report T-5-2023, Los Alamos Scientific Laboratory, 1957.

11. COCHRAN, J. A., *The Analysis of Linear Integral Equations*, McGraw-Hill, New York, 1972.

12. CORMACK, A. M., *Representation of a function by its line integrals with some radiological applications*, J. Appl. Phys., 34 (1963), pp. 2722–2727.

13. —, *Representation of a function by its line integrals with some radiological applications, II*, J. Appl. Phys., 35 (1964), pp. 2908–2913.

14. DEANS, S. R., *The Radon Transform and Some of Its Applications*, John Wiley, New York, London, 1983.

15. DELVES, L. M., and MOHAMED, J. L., *Computational Methods for Integral Equations*, Cambridge Press, Cambridge, 1985.

16. DELVES, L. M., and WALSH, J., *Numerical Solution of Integral Equations*, Clarendon Press, Oxford, 1974.

17. DIAZ, J., AND METCALF, F. T., *On iterative procedures for equations of the first kind, $Ax = y$, and Picard's criterion for the existence of a solution*, Math. Comp., 24 (1970), pp. 923–935.

18. DONGARRA, J. J., BUNCH, J. R., MOLER, C. B., AND STEWART, G. W., *LINPACK Users' Guide*, Society for Industrial and Applied Mathematics, Philadelphia, 1979.

19. ENGL, H. W., and GFRERER, H., *A posteriori parameter choice for general regularization methods for solving ill-posed problems*, Appl. Numer. Math., 4 (1988), pp. 395–417.

20. FRANKLIN, J., *Matrix Theory*, Prentice-Hall, Englewood, Cliffs, NJ, 1968.

21. —, *Well-posed stochastic extensions of ill-posed linear problems*, J. Math. Anal. Appl., 31 (1970), pp. 682–716.

22. GOLBERG, M. A., *A method of adjoints for solving some ill-posed equations of the first kind*, Appl. Math. Comp., 5 (1979), pp. 123–130.

23. GROETSCH, C. W., *Comments on Morozov's discrepancy principle*, in Improperly Posed Problems and their Numerical Treatment, Hämmerlin, G. and Hoffman, K.-H., eds., Birkhäuser, Basel, 1983.

24. —, *The Theory of Tikhonov Regularization for Fredholm Integral Equations of the First Kind*, Pitman, Boston, 1984.

25. GROETSCH, C. W., and NEUBAUER, A., *Regularization of ill-posed problems: Optimal parameter choice in finite dimensions*, J. Approx. Theory, 58 (1989), pp. 184–200.

26. HADAMARD, J., *Le problème de Cauchy et les équations aux dérivée partielle linéaires hyperboliques*, Hermann, Paris, 1932.

27. HANSON, R. J., *A numerical method for solving integral equations of the first kind using singular values*, SIAM J. Numer. Anal., 8 (1971), pp. 616–622.

References

28. HARDY, G. H., LITTLEWOOD, J. E., AND PÓLYA, G., *Inequalities*, Cambridge University Press, London, 1934.

29. HENDRY, W. L., *A Volterra integral equation of the first kind*, J. Math. Anal. Appl., 54 (1976), pp. 266–278.

30. HERMAN, G. T., ed., *Image Reconstruction from Projections*, Springer-Verlag, Berlin, 1979.

31. de HOOG, F. R., *Review of Fredholm integral equations of the first kind*, in The Application and Numerical Solution of Integral Equations, Anderssen, R. S., et al., eds., Sijthoff and Noordhoff, the Netherlands, 1980.

32. KRALL, A. M., *Applied Analysis*, D. Reidel, Dordrecht, Boston, 1986.

33. KRESS, R., *Linear Integral Equations*, Springer-Verlag, Berlin, 1989.

34. LANDWEBER, L., *An iteration formula for Fredholm integral equations of the first kind*, Amer. J. Math., 73 (1951), pp. 615–624.

35. LOUIS, A. K., and MAASS, P., *A mollifier method for linear operator equations of the first kind*, Inverse Problems, 6 (1990), pp. 427–440.

36. MACRAE, R., AND WING, G. M., *Some remarks on problems connected with data analysis in PHERMEX*, Technical Report T-5-2043, Los Alamos Scientific Laboratory, 1958.

37. MCLAUGHLIN, D. W., ed., *Inverse Problems*, American Mathematical Society, Providence, 1984.

38. MIYOMOTO, S., AND SAVAROGI, Y., *Identification of distributed systems and the theory of regularization*, J. Math. Anal. Appl., 63 (1978), pp. 77–95.

39. MOROZOV, V. A., *Methods for Solving Incorrectly Posed Problems*, Springer-Verlag, New York, 1984.

40. NATTERER, F., *The Mathematics of Computerized Tomography*, John Wiley, New York, London, 1986.

41. PHILLIPS, D. L., *A technique for the numerical solution of certain integral equations of the first kind*, J. Assoc. Comput. Mach., 9 (1962), pp. 84–97.

42. PRESS, W., FLANNERY, B., TEUKOLSKY, S., AND VETTERLING, W., *Numerical Recipes*, Cambridge University Press, Cambridge, 1986.

43. RIESZ, F., AND SZ-NAGY, B., *Functional Analysis*, Frederick Ungar, New York, 1955.

44. SHAW, C. B. Jr., *Improvement of the resolution of an instrument by numerical integration of an integral equation*, J. Math. Anal. Appl., 37 (1972), pp. 83–112.

45. SHEPP, L. A., ed., *Computed Tomography*, Proc. of Symposium in Applied Mathematics, American Mathematical Society, Providence, 1983.

46. SMITHIES, F., *Integral Equations*, Cambridge University Press, London, 1958.

47. STEWART, G. W., *Introduction to Matrix Computation*, Academic Press, New York, 1973.

48. STRAND, O. N., *Theory and methods related to the singular function expansion and Landweber's iteration for integral equations of the first kind*, SIAM J. Numer. Anal., 11 (1974), pp. 798–825.

49. TAYLOR, A. E., *Introduction to Functional Analysis,* John Wiley, New York, London, 1958.

50. TIKHONOV, A. N., *On the solution of ill-posed problems and the method of regularization*, Soviet Math. Dokl., 4 (1963), pp. 1035–1038.

51. TIKHONOV, A. N., AND AVSENIN, V. Y., *Solution of Ill-posed Problems*, Winston, Washington, DC, 1977.

52. TIKHONOV, A. N., AND GONCHARSKY, A. V., *Ill-posed Problems in the Natural Sciences,* translated by M. Bloch, MIR Publishers, Moscow, 1987.

53. TWOMEY, S., *On the numerical solution of Fredholm integral equations of the first kind by the inversion of the linear system produced by quadrature,* J. Assoc. Comput. Mach., 10 (1963), pp. 97–101.

54. WAHBA, G., *Smoothing and ill-posed problems*, in Solution Methods for Integral Equations, M. Golberg, ed., Plenum, New York, 1979.

55. WHITTACKER, E. T., AND WATSON, G. N., *Modern Analysis*, Cambridge University Press, London, 1943.

56. WIDDER, D., *The Laplace Transform,* Princeton University Press, Princeton, 1946.

57. WING, G. M., *An Introduction to Transport Theory*, John Wiley, New York, London, 1962.

Appendix A: The Domain of $K(x, y)$

We have found it convenient to consider the kernel $K(x, y)$ to be defined on a square $a \leq x, y \leq b$. In a physical context the validity of this assumption is not always clear. For example, x might be a spatial variable whereas y could stand for energy. (See Section 2.2.) The two variables are not even comparable. How can we suppose that the point (x, y) lies in a square?

There are many ways to approach this question. I present one. Suppose that in context the meaningful values of x lie in $c \leq x \leq d$ while y lies in $a \leq y \leq b$. I consider these numbers to be dimensionless. Define

$$z = \frac{b-a}{d-c}x + \frac{ad-bc}{d-c} \quad \text{or} \quad x = \frac{d-c}{b-a}z + \frac{bc-ad}{b-a}.$$

Then as x ranges from c to d, z varies between a and b.

Consider the question of IFKs, and suppose that

$$g(x) = \int_a^b K(x,y)f(y)dy, \quad c \leq x \leq d, \quad a \leq y \leq b.$$

Rewriting,

$$g\left(\frac{d-c}{b-a}z + \frac{bc-ad}{b-a}\right) = \int_a^b K\left(\frac{d-c}{b-a}z + \frac{bc-ad}{b-a}, y\right) f(y)dy.$$

Set

$$\tilde{g}(z) = g\left(\frac{d-c}{b-a}z + \frac{bc-ad}{b-a}\right),$$

$$\tilde{K}(z,y) = K\left(\frac{d-c}{b-a}z + \frac{bc-ad}{b-a}, y\right)$$

and observe that
$$\tilde{g}(z) = \int_a^b \tilde{K}(z,y)f(y)dy,$$
$$a \leq z, y \leq b$$
as desired.

Appendix B: Remarks about Complex Functions, Vectors, and Operators

The assumption is made throughout this work that all quantities involved are real. Extension to the complex case is quite easy.

First suppose that $h_1(x)$ is a complex function. Its sup-norm, one-norm, and two-norm can all be defined exactly as in Section 3.2. If $\mathbf{h}_1$ is a complex vector, its norms may be defined as in Section 3.3.

If $h_1(x)$ and $h_2(x)$ are both complex, it is convenient to define $(h_1, h_2) = \int_a^b h_1(x)\overline{h_2(x)}dx$. We observe that with this definition $(h_1, h_2) = \overline{(h_2, h_1)}$ and $(h_1, h_1) = \| h_1 \|_2^2$. For complex vectors $\mathbf{h}_1$ and $\mathbf{h}_2$ we set

$$(\mathbf{h}_1, \mathbf{h}_2) = \sum_{j=1}^n h_{1j}\overline{h}_{2j},$$

where h_{1j} and h_{2j} are the respective components of $\mathbf{h}_1$ and $\mathbf{h}_2$.

Now suppose the operator K is associated with the complex kernel $K(x, y)$. Then K^* is represented by the kernel $\overline{K}(y, x)$. An elementary calculation shows that $(Kh_1, h_2) = (h_1, K^*h_2)$, where the inner product used is that above. Note that $K^* = K$ if and only if $K(x, y) = \overline{K}(y, x)$. This generalizes the idea of symmetry. An operator that has the property that $K = K^*$ is called *Hermitian*. In the complex case, Hermitian operators enjoy the same properties that symmetric real operators do.

It is important to notice that a complex kernel can be symmetric without being Hermitian, and many desirable properties are lost. For example, the kernel $e^{i|x-y|}$ is in this category.

Appendix C: Modes of Convergence

Reference has been made in the text to the fact that convergence (usually of infinite series) must always be properly interpreted. I give a very brief discussion of that point.

Recall the ordinary definition of convergence of a series of constants: the infinite series $\sum_{j=1}^{\infty} a_j$ converges to the sum S if $\lim_{n\to\infty} S_n = S$, where $S_n = \sum_{j=1}^{n} a_j$.

The question of convergence of a series of functions reduces immediately to the above definition provided we are interested in a specific value of the independent variable. The infinite series $\sum_{j=1}^{\infty} f_j(x)$ converges at the point x' to the sum $S(x')$ if $\lim_{n\to\infty} S_n(x') = S(x')$, where $S_n(x) = \sum_{j=1}^{n} f_j(x)$.

If we are interested in the convergence of a series of functions over an interval $a \leq x \leq b$, the concept of uniform convergence is often useful. The infinite series $\sum_{j=1}^{\infty} f_j(x)$ converges *uniformly* on $a \leq x \leq b$ if $\lim_{n\to\infty} S_n(x) = S(x)$ uniformly, $S_n(x)$ as above. I refer you to any advanced calculus text for the usual definition of uniform convergence of a sequence.

There is a (probably) unfamiliar equivalent definition of uniform convergence. Let the sup-norm be defined as usual:

$$\| g \|_S = \sup_{a \leq x \leq b} | g(x) | .$$

Then the infinite series $\sum_{j=1}^{\infty} f_j(x)$ converges *uniformly* for $a \leq x \leq b$ if and only if

$$\lim_{n\to\infty} \| S_n - S \|_S = 0.$$

This way of formulating uniform convergence suggests that it might be interesting to replace the sup-norm by other norms. Thus we can say that the series converges *in norm* to S if

$$\lim_{n \to \infty} \| S_n - S \| = 0.$$

Although it may seem that this notion is introduced only to be esoteric, it turns out that the concept of convergence in norm is an important one. In fact, many of the results stated imprecisely in this book become rigorously correct if we interpret convergence to be in the *two-norm*. See Section 7.2.

The question immediately arises, "Are not all the modes of convergence equivalent?" By no means. Consider, for example,

$$S_n(x) = n^2 x e^{-nx}.$$

(Because convergence of infinite series really depends upon convergence of a sequence of partial sums, there is no reason to start with the series.) It is easy to see that for $0 \leq x \leq 1$,

$$\lim_{n \to \infty} S_n(x) = S(x) = 0.$$

This convergence is *not* uniform. Moreover,

$$\int_0^1 | S_n(x) - S(x) |^2 \, dx = \int_0^1 n^4 x^2 e^{-2nx} dx = n \int_0^n t^2 e^{-2t} dt \to \infty,$$

and so the sequences fails to converge to zero in the two-norm. (It also fails in the one-norm.)

It is also possible to have convergence in the one- or two-norm without having ordinary (pointwise) convergence. I construct a rather artificial example. Define

$$\tilde{S}_1(x) = 1, \quad 0 \leq x \leq \frac{1}{2}, \quad \tilde{S}_1(x) = 0 \text{ otherwise};$$

$$\tilde{S}_2(x) = 1, \quad \frac{1}{2} \leq x \leq 1, \quad \tilde{S}_2(x) = 0 \text{ otherwise};$$

$$\tilde{S}_3(x) = 1, \quad 0 \leq x \leq \frac{1}{4}, \quad \tilde{S}_3(x) = 0 \text{ otherwise};$$

$$\tilde{S}_4(x) = 1, \quad \frac{1}{4} \leq x \leq \frac{1}{2}, \quad \tilde{S}_4(x) = 0 \text{ otherwise};$$

Appendix C

$$\tilde{S}_6(x) = 1, \quad \frac{3}{4} \leq x \leq 1, \quad \tilde{S}_6(x) = 0 \text{ otherwise};$$

$$\tilde{S}_7(x) = 1, \quad 0 \leq x \leq \frac{1}{8}, \quad \tilde{S}_7(x) = 0 \text{ otherwise};$$

$$\tilde{S}_8(x) = 1, \quad \frac{1}{8} \leq x \leq \frac{1}{4}, \quad \tilde{S}_8(x) = 0 \text{ otherwise};$$

....

$$\tilde{S}_{14}(x) = 1, \quad \frac{7}{8} \leq x \leq 1, \quad \tilde{S}_{14}(x) = 0 \text{ otherwise};$$

etc.

Although it is possible to write an expression for the general term, I shall not bother. Now it is easy to see geometrically that

$$\lim_{n \to \infty} \int_0^1 \tilde{S}_n^2(x) dx = 0.$$

Thus

$$\lim_{n \to \infty} \| \tilde{S}_n - 0 \|_2 = 0,$$

so that $\tilde{S}_n$ has limit zero in the two-norm (and also in the one-norm). However, this sequence does not converge at any point in $0 \leq x \leq 1$. To understand this, pick a specific point x' and notice that for infinitely many n's $S_n(x') = 1$ and for infinitely many more n's $S_n(x') = 0$.

The overall question of which modes of convergence imply which others is a fascinating one that cannot be pursued here. For our purpose it suffices to know that because of our restriction that all functions with which we deal are "nice," it is *usually* acceptable to think in terms of ordinary convergence. In many cases this is difficult to prove rigorously. Strangely enough, convergence in the one- or two-norm is often much easier to verify than ordinary convergence. That is one of the reasons for the importance of the concept.

Index

Abel equation, 12, 20, 23
absorption, 9
adjoint operator, 62
algorithm, 8, 72, 114
analytic function, 8

B-splines, 88
Bessel inequality, 67

Chebyshev polynomials, 14, 24
codes, 2, 72, 114
collocation, 87
compact operator, viii
complex quantities, 114, 127
condition number, 79, 117
constraints, 96
continuity, 3
convergence of series, 66, 129
    in two-norm sense, 130
    modes of, 58, 130
    uniform, 129
convolution (Faltung), 2, 23, 69
correlation, 104
cross section, 9, 11

data, 3, 74, 115
deconvolution, vii, 2
density, 11
detector, 9
Diaz, J., 106
differential equation, 21
differential operator, 32, 34
Dirac delta function, 3, 77, 88, 105, 114, 116
discretization, 59
domain, 125

dual, 61

eigenfunction, 45, 56, 61, 89
eigenvalue of integral operator, 45, 56
eigenvalue of matrix, 56, 59
eigenvector of matrix, 45
energy, 9, 15
    spectrum, 10
errors, 3
    in data 12, 19, 21, 74, 79
    in kernel, 4, 117
Euclidean length, 29
Euler, 110
expected value, 104

finite-dimensional operator, 36
filter, 107
fish population, 18, 24
fission, 9, 11
folding, 2
Fourier series, 13, 42
Franklin, J., 104
Fredholm, 1, 65
functional 93, 96, 101
functional analysis, 27, 38, 116

Galerkin, B. G., 88, 108
Gaussian, 91, 116
geomagnetic prospecting, 17
Golberg, M., 115
Grammian, 44
Green's function, 5
ground rules, 3, 113

Hadamard, J., 21

133

Index

Hendry, W., 16
Hermitian, 127
histogram, 73

identity, 47, 105
IFK, 1
ill-posed problem, 21, 56
image restoration, 4
integral equation, vii
    first kind, vii, 1, 3
    Fredholm, 1
    linear, 1
    nonlinear, 4, 18
    second kind, 1, 46, 63
    Volterra, 59
inverse, 52
    of integral operator, 32, 52
    of matrices, 73
isotropic distribution, 15
iterative methods, 106

kernel, 1
    approximate, 80
    errors in, 117
    flatness of, 77, 79, 119
    general, 51, 65
    post-pile, 80, 86
    separable (degenerate) 41, 63, 80
    symmetric, 46, 56, 63

Landweber, L., 106
Laplace transform, 7, 10, 23, 69
    inverse of, 7, 8, 24, 37
least squares, 81
Lebesgue integral, viii, 113
linear algebra, 1
linear independence, 41
linear operator, 52

matrices, 35, 63, 72
    norms of, 35
    theory of, 1, 41, 56
mesh size, 103
Metcalf, F. T., 106
moments, 115
mollifier, 115

Neumann series, 70
nonlinear IFKs, 4, 18
norm, of vector or function, 27
    Euclidean, 29
    general, 28
    one- (or L_1-), 28, 29
    sup, 27, 29
    two- (or L_2-), 28, 29
norm, of operator, 36
    differential, 32, 35
    general, 36
    integral, 30, 31
    matrix, 35
normed linear function space, 28, 34
numerical differentiation, 12, 22, 27
numerical methods, 2, 37, 72

operators, 30, 32, 35, 36
    bounded, 30, 36
    inverse, 52
    linear, 52
    nonsymmetric, 58
    symmetric, 60
    unbounded, 37, 75
    Volterra, 59

orthogonal functions, 14, 57, 87

parameterization, 91
PHERMEX, 11, 13
Phillips, D. L., 92
Post–Widder formula, 8, 12
probability, 9, 22
projection, 87

quadrature, 49, 72, 73, 83, 85, 114

random variable, 74, 104
range, 58
regularization, 92, 97, 103
remote sensing, 2
resolve, ix, 2, 71
Riemann-Lebesgue theorem, 55, 67
rod model, 15
root mean square, 75
roughening operator, 34

scattering, 9, 11
Schwarz inequality, 39
series expansions, 85, 89, 100
signal, 9
singular functions and vectors, 58, 89
    of integral operators, 61
    of matrix operators, 63

Index

singular value, 61
    decomposition, 63, 80, 90, 114
    of integral operators, 61
    of matrix operators, 63
smoothing operator, 31
source strength, 15
space, 28, 34, 54
statistical approach, 103
stochastic approach, 103
Strand, O., 107
supremum, 27
symmetric kernel, 46
symmetric matrix, 46
symmetric operator, 56

tautochrone, 19
Tichonov, A. N., 92, 103, 107
tomography, 13
transport theory, 8
Twomey, S., 92

unfold, vii, 2
unique solution, 21, 43, 47
unstable, 21

Volterra, V., 12, 59, 69, 70

Wahba, G., 105
well-posed problem, 21
white noise, 105

x ray, 9, 11